AF718060

THEMES OF THE Times

ON

Astronomy

A COLLECTION OF ARTICLES FROM

The New York Times

PEARSON
Addison Wesley

San Francisco Boston New York
Cape Town Hong Kong London Madrid Mexico City
Montreal Munich Paris Singapore Sydney Tokyo Toronto

Editor-in-Chief: Adam Black
Associate Editor: Stacie Kent
Executive Marketing Manager: Christy Lawrence
Managing Editor, Production: Erin Gregg
Production Supervisor: Shannon Tozier
Production Assistant: Benjamin Lau
Production Management and Composition: TechBooks/GTS
Manufacturing Buyer: Michael Early
Text and Cover Printer: Courier, Stoughton

All pedagogical content, including article summaries and "In Review" article-related questions, was created by:

Doug Lombardi, University of Arizona
Jeffrey Bennett, University of Colorado
Megan Donahue, Michigan State University
Nicholas Schneider, University of Colorado
Mark Voit, Michigan State University

ISBN 0-8053-9028-6

Articles copyright 1999, 2000, 2002, 2003, 2004 The New York Times Company. Reprinted with permission.

All pedagogical content for this supplement has been created by Pearson Addison-Wesley, and includes: article summaries; cross-references to *The Cosmic Perspective*, Third Edition, *The Essential Cosmic Perspective*, Third Edition, and *The Cosmic Perspective*, Fourth Edition, in the Table of Contents; and "In Review" questions that follow each article.

Copyright ©2006 Pearson Education, Inc., publishing as Addison Wesley, 1301 Sansome St., San Francisco, CA 94111. All rights reserved. Manufactured in the United States of America. This publication is protected by Copyright and permission should be obtained from the publisher prior to any prohibited reproduction, storage in a retrieval system, or transmission in any form or by any means, electronic, mechanical, photocopying, recording, or likewise. To obtain permission(s) to use material from this work, please submit a written request to Pearson Education, Inc., Permissions Department, 1900 E. Lake Ave., Glenview, IL 60025. For information regarding permissions, call (847) 486-2635.

Many of the designations used by manufacturers and sellers to distinguish their products are claimed as trademarks. Where those designations appear in this book, and the publisher was aware of a trademark claim, the designations have been printed in initial caps or all caps.

PEARSON
Addison Wesley

1 2 3 4 5 6 7 8 9 10 -CRS- 08 07 06 05
www.aw-bc.com/physics

THEMES OF THE Times ON Astronomy

Contents

to accompany chapters of *The Cosmic Perspective*, Third Edition

THEMES OF THE Times ON Astronomy

Contents

to accompany chapters of *The Essential Cosmic Perspective*, Third Edition

THEMES OF THE Times ON Astronomy

Contents

to accompany chapters of *The Cosmic Perspective*, Fourth Edition

With the exception of the Moon, the planet Venus is the brightest object in the night sky. Shining brightly just after sunset or before sunrise, Venus has captured the imagination of sky watchers since ancient times. As a planet, Venus wanders across the sky in a manner much different than the stars, and occasionally Venus appears to move across the Sun. Because Venus is much more distant from the Earth than the Moon is, when its path transverses the solar disk it does not fully eclipse the light, but rather is seen (through special filtered safety glass) as a small blemish. This movement, called a solar transit, is more than a fanciful curiosity. Such transits of Venus have helped scientists accurately determine planetary distances.

Venus Returns for Its Shining Hour

By Warren E. Leary
***The New York Times*, May 18, 2004**

The world is about to witness a rare spectacle that once launched expeditions to ideal vantage points around the globe and inspired millions of people to venture outside and stare at the heavens.

On June 8, people in the right places on Earth will be able to see Venus move across the face of the Sun in a kind of minieclipse that is visible twice every century or so. The last such occurrence, called a transit of Venus, was in 1882. It inspired an international effort to use the event to answer one of the most pressing scientific questions of the day: What is the exact distance between the Sun and Earth?

Although studies of the event failed to provide an exact answer, they did narrow the range of estimates and measurements, and ushered in an era of investing in science as a symbol of national prestige. For the last event, the United States government mustered eight expeditions to make observations around the world, partly because Britain, France, Russia and other rivals did the same.

By bouncing radar signals off the Sun and Venus and using spacecraft measurements, scientists in the 1960's calculated that the average Sun-to-Earth distance is 92,955,859 miles, a measure called the astronomical unit.

Scientists realized for centuries that if they could find out that number, they could use the formulas of the 17th-century astronomer Johannes Kepler to calculate the size of the solar system and the exact distances between the planets.

"This was the most important question of its day in astronomy," said Dr. Jay M. Pasachoff, a professor of astronomy at Williams College. "And using the transits of Venus to calculate the astronomical unit was the best way to do it."

Although transits of Venus have occurred for thousands of years, the first report of its subtle crossing of the Sun was in 1639. The transits occur when the orbits of Venus, Earth and the Sun put them into alignment along the same plane.

Since 1639, transits have occurred in 1761, 1769, 1874 and 1882. If someone misses the one next month, there will be another opportunity on June 6, 2012. After that, more than a century will pass before the next transits, in 2117 and 2125. Because of its rarity, the transit next month, best viewed from Europe and the Mideast, is generating great scientific and public interest, said Dr. Steven J. Dick, chief historian for the National Aeronautics and Space Administration. Dr. Dick has written extensively on the 18th- and 19th-century transits.

No one alive today saw the last transit, he said, and seeing the next two will be the only chance most people have.

"These are truly once-in-a-lifetime events," Dr. Dick said. "Although the scientific importance has diminished, I think there will be a lot of interest this time among the public, based on e-mail I've seen from around the world."

Dr. David DeVorkin, curator of the history of astronomy at the National Air and Space Museum, said the 1874 and 1882 transits were prominently featured in newspapers and magazines. A carnival atmosphere pervaded Wall Street for the transit on Dec. 6, 1882, with people crowding the area and staring up through smoked glass.

"It was a popular diversion," Dr. DeVorkin said. "Something maybe everybody didn't try to see, but everybody talked about it."

Scientific interest persists. Instruments aboard at least three Sun-watching satellites, as well as ground telescopes, will follow the event. Researchers will use Venus' trek to test techniques and instruments that can be used to detect planets in other solar systems.

More than 120 extrasolar planets have been discovered orbiting other stars, most of them huge bodies found because their gravity affected the motion of their stars.

Astronomers have recently detected a small number of far planets by measuring the fluctuations that they cause in light from the stars they circle. In 2007, NASA plans to launch the Kepler spacecraft to monitor Sun-like stars in hope of detecting Earth-size planets through small decreases in star brightness.

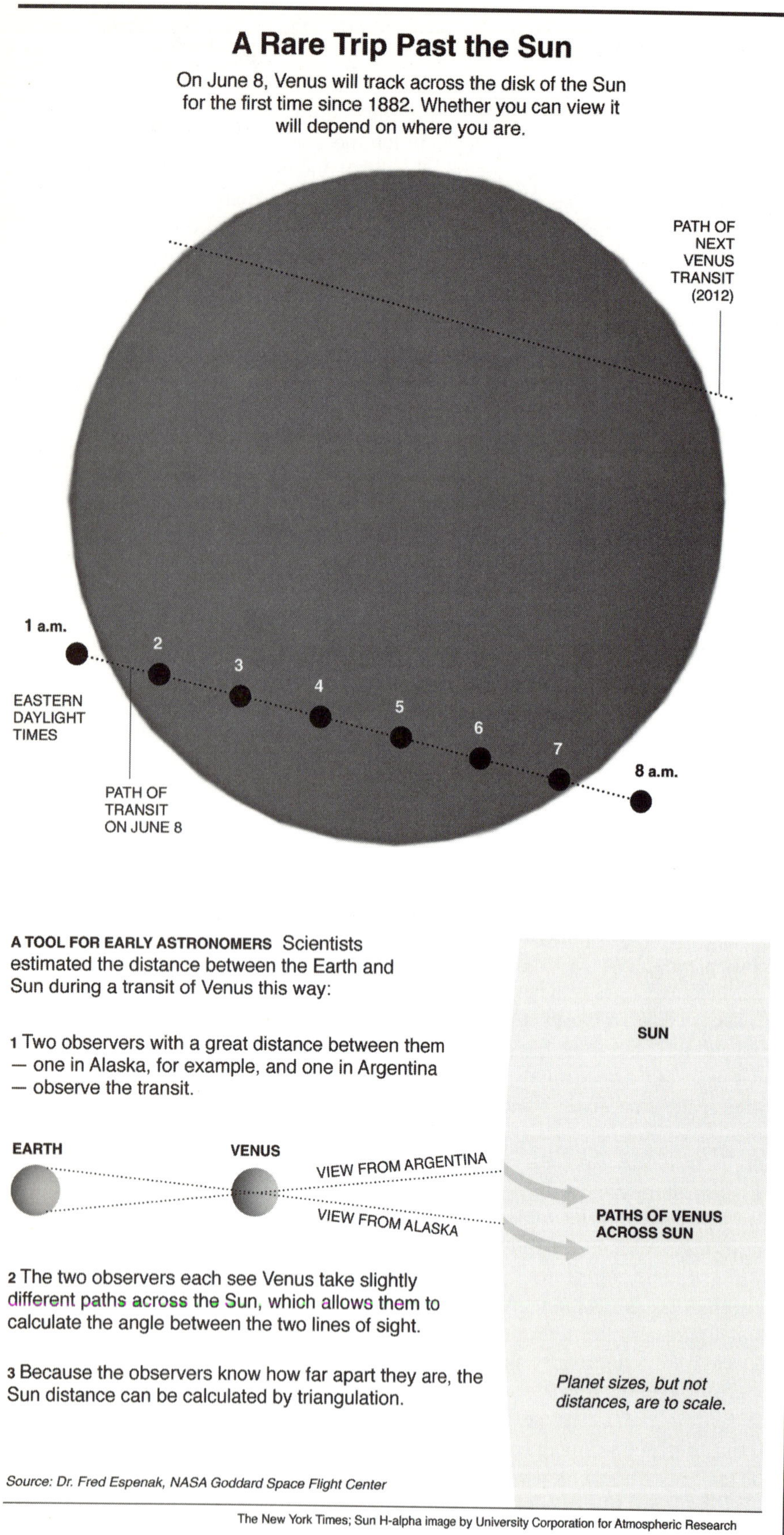

Although denied a direct view of the transit because it occurs at night in the American West, astronomers with the University of Arizona hope to get an indirect view. Dr. Glenn H. Schneider said he and a colleague, Paul S. Smith, would try to use the Steward Observatory in Tucson to measure about a half-hour of sunlight from the end of the transit as it reflects off the Moon.

"We want to see if we can detect the signature of Venus' atmosphere spectroscopically from sunlight reflecting off the moon, as if it was a reading coming from a faraway star," Dr. Schneider said.

The transits generally occur in a predictable pattern of two occurring in an eight-year period, followed by one 105½ years later and another eight years after that. After an additional 121½ years, the pattern repeats. The paired eight-year sightings occur because a Venusian year equals 224.7 Earth days, making 13 Venusian years equal to eight Earth years.

That allows the planets to return to about the same alignment with the Sun they had been in eight years earlier, after which they go out of sync for more than a century.

On Tuesday, June 8, observers lucky enough to view the entire transit will see Venus as a small black spot crossing the southern hemisphere of the Sun from left to right. The planet, entering the disc of the Sun at the 8 o'clock position, will take six hours to cross the bright face before exiting at the 5 o'clock position.

Venus, appearing as a round black dot with a diameter one thirty-second of the Sun's, is widely expected to cause a one-tenth of 1 percent drop in sunlight that reaches Earth during the event.

Location is everything, particularly when trying to witness celestial events. The entire transit will be visible in Europe, most of Africa, the Mideast and most of Asia. The unlucky regions of the globe where the event occurs at night, and is unviewable, include western North America, including most of the United States west of the Rockies;

southern Chile and Argentina; Hawaii; and New Zealand.

Some regions will see just part of the transit, because the Sun sets while it is in progress. Those areas include Australia, Indonesia, Japan, the Philippines, Korea and Southeast Asia.

Likewise, the Sun rises with the transit in progress over eastern North America, the Caribbean, western Africa and most of South America, allowing observers a brief view before the event ends. How much early risers see will depend on the weather and how high the Sun rises above the horizons before Venus moves out of view.

In New York, sunrise will be at 5:25 a.m., and Venus is to begin exiting the solar disc at 7:06, when the Sun is 17 degrees above the horizon. The planet's final contact with the edge of the Sun should occur at 7:26 a.m., when the Sun is 20 degrees high. Times are similar for most cities in the Eastern time zone and one hour earlier in the Central time zone. But moving West means that the Sun is lower on the horizon.

Modern interest in planetary transits can be traced from Kepler. Based on his calculations of planetary motion, he wrote in 1627 that Mercury would cross the face of the Sun in November 1631 and that Venus would follow on Dec. 6 that year. Kepler suggested that observers placed at widely different points on Earth could indirectly calculate the distance to the Sun by using Venus.

Knowing the distance between observers and the different angles from which they viewed the transit, astronomers could calculate the distance to Venus and use that to compute the Earth-to-Sun measurement, he reasoned.

Kepler died the year before the 1631 Venus transit, but he would not have seen it had he lived, because it occurred at night in Europe. He would have also missed the next transit, in 1639, because he made a miscalculation that failed to predict it.

Fortunately, a young English astronomer, Jeremiah Horrocks, became

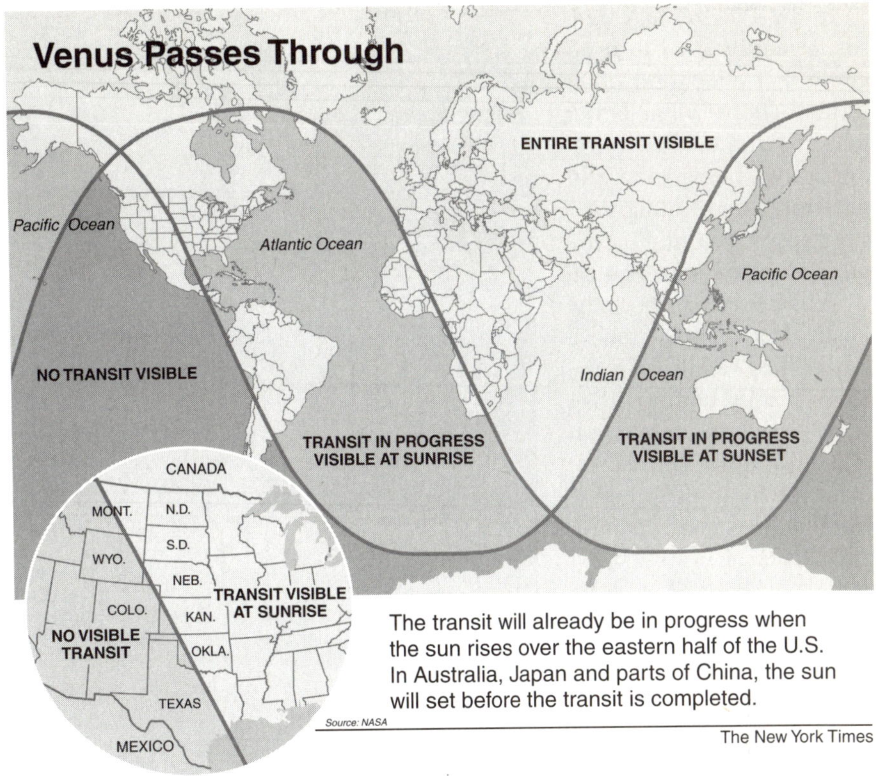

The transit will already be in progress when the sun rises over the eastern half of the U.S. In Australia, Japan and parts of China, the sun will set before the transit is completed.

Source: NASA

The New York Times

interested in Kepler's work and, in recalculating some of the German's tables, discovered that a transit would occur on Nov. 24, 1639. Horrocks witnessed part of the transit from his home in Much Hoole, Lancashire, and a friend whom he notified by letter, William Crabtree, saw it from Manchester.

The next transits, in the 18th century, drew much more attention, thanks to Edmond Halley, the British astronomer best known for the comet that bears his name. Halley suggested using the 1761 and 1769 transits to calculate the Sun-to-Earth distance by having observers time the events from widely spaced latitudes and trace the planet's path across the Sun's face as they saw it from their positions. By measuring the angular shifts of the paths based on the timings, Halley reasoned, the astronomical unit could be calculated.

Although Halley died in 1742, his plan guided many observations made of the two transits from around the world. But the results varied widely and were disappointing. Among those trying to work on the problem in 1769 was the British explorer Capt. James Cook, who took his ship, the Endeavour, on its first voyage to the South Pacific to observe the transit from Tahiti.

Cook and others were frustrated in their observations by the inability to time the exact moment when the edges of the planet and the Sun appeared to touch. When Venus nears the edge of the disc of the Sun, its black circle appears to ooze toward the edge of the sun without showing a clear point of contact. Although the precise second of contact was needed for calculations, this so-called "black drop" phenomenon caused observers watching the same event to disagree by several seconds up to a minute on when the outer edges touched.

Cook and other observers speculated that the problem was the distortion of light through the Venusian atmosphere.

Earlier this year, using spacecraft observations, Dr. Pasachoff and other scientists concluded that the black drop effect was caused by a combination of images' blurring in small-aperture telescopes and the natural dimming of sunlight near the Sun's visible edge.

In the 19th-century transits, scientists tried to overcome that effect and other imperfections with better telescopes and the introduction of photography. Still measuring and timing transits never led to finding the precise Sun-to-Earth distance.

William Harkness of the United States Naval Observatory refined results from the 1882 transits and in 1894 came up with an astronomical unit measure of 92,797,000 miles. But the work of another Naval Observatory scientist, Simon Newcomb, was adopted as the world standard at a 1896 meeting in Paris, Dr. Dick said. Newcomb, who gave little credence to transit data, combined values from several sources including speed-of-light star readings, to come up with a figure of about 92,872,000 miles. Both were close to the modern value of 92,955,859 miles, but precision is critical in astronomical terms.

Photo by U.S. Naval Observatory

Testing the equipment at the 1874 American transit-of-Venus observation station on Kerguelen Island. On June 8, and for the first time since 1882, Earthlings will stare at the heavens with a chance to see Venus move across the face of the Sun.

Nevertheless, Dr. Dick said, the transits of Venus remain important because the desire to define the astronomical unit—and to maintain or gain scientific prestige—led many nations to mount competing expeditions. In 1874, Russian sent out 26 expeditions, Britain 12, the United States 8, Germany and France 6 each, Italy 3 and the Netherlands 1.

"You could compare it with the space race in the 20th century," he said.

How to Watch Without Harm

When viewing solar events like eclipses or the transit of Venus, precautions are needed. Never look directly at the Sun. The direct gaze can lead to severe eye damage or blindness, experts say.

Sunglasses and clouds do not protect the eyes, and viewing the Sun through unfiltered telescopes, binoculars or cameras can result in instant and permanent damage. Telescopes and binoculars should be equipped with special undamaged solar filters. Glasses with solar lenses are available commercially, but even then do not stare at the Sun for long periods.

The transit of Venus can safely be seen if viewed indirectly, using telescopes or pinhole boxes to focus the image on a screen or paper opposite the Sun.

Information on viewing solar events is at these Web sites:

www.transitofvenus.org/safety.htm
http://sunearth.gsfc.nasa.gov/eclipse/SEhelp/safety2.html

More Web resources can be found at: nytimes.com/science.

IN REVIEW

1. In what ways did the 1882 Venus transit usher in a new era of using science to bolster national prestige?
2. Why would two transits that are just seven years apart be advantageous for calibrating the distance between the Earth and the Sun compared to transits that are spaced more than one hundred years apart?
3. We can now accurately measure the distance from the Earth to the Sun using radar. But the Venus transit still provides some useful information to astronomers. List some different types of information that scientists get from the transit.
4. Before radar was invented, how did astronomers use transits to calculate the distance from the Earth to the Sun?
5. Why would Mercury and Venus be the only transits visible from Earth? If you were standing on the surface of Saturn's moon Titan, what planetary transits might you be able to see?

Ancient Greek scientists, even without telescopes and computers, laid the foundation for our understanding of the cosmos. But understanding the exact nature of individual contributions is very difficult because two to three thousand years have passed. Much like the fictional Indiana Jones, modern scientists and historians search for clues in museum antiquities and can come up with some amazing discoveries by drawing on a deep understanding of both astronomy and history. In essence, our modern and precise understanding of how the Earth wobbles and how the positions of the constellations change with time can tell us how Greek scientists viewed the night sky so long ago. Only because of their painstakingly precise record keeping can we make sense of what their contributions might have been.

Visions of Ancient Night Sky Were Hiding in Plain Sight for Centuries

By Kenneth Chang
***The New York Times*, January 18, 2005**

Two thousand years later, astronomers still talk about Hipparchus' star catalog, the earliest known compendium of the night sky.

Hipparchus, who lived in Greece during the second century B.C., was perhaps the world's first great astronomer. He calculated, within six and a half minutes, the length of a year. He figured out that Earth's axis wobbles as it spins. For his star catalog, completed in 129 B.C., he devised a coordinate system to plot each star's location and a scale to rank the brightness. Astronomers still use this magnitude scale today.

Most of Hipparchus' work, however, is known only secondhand. No one has seen the catalog for centuries, a fabled ancient text apparently lost forever.

No one until Dr. Bradley E. Schaefer thought of looking for it on a statue.

At a meeting of the American Astronomical Society here, Dr. Schaefer, a professor of physics at Louisiana State University, reported that the text of Hipparchus' catalog was still lost, but that he had found possibly the next best thing, a pictorial representation of the catalog's contents.

"Here we have a real example of lost ancient knowledge being discovered," Dr. Schaefer said. Perhaps not quite the da Vinci code, but, as he described it, "some of the most influential lost ancient lore."

Dr. Owen Gingerich, an expert on the history of astronomy at Harvard, described Dr. Schaefer's research as "quite astonishing."

Remarkably, the new discovery has been in plain view for centuries.

A seven-foot-tall statue at the National Archaeological Museum in Naples, Italy, shows the god Atlas kneeling with a globe weighing on his shoulders. Atlas was sentenced by Zeus to hold up the sky, and the globe is a depiction of the night sky as seen from Earth, with pictures of Aries the ram, Cygnus the swan, Hercules the hero and other animals and people representing 41 constellations. The statue, known as the Farnese Atlas, is the oldest surviving pictorial record of Western constellations. It dates to Roman times, around A.D. 150, and is known to be a copy of an earlier Greek work.

"It was just sitting there, waiting," Dr. Schaefer said.

Dr. Edwin C. Krupp, director of the Griffith Observatory in Los Angeles, said modern notions tended to discount Greek myths as more astrology than astronomy. "We tend to think they were silly stories that the ancients told," he said. "I don't think anyone thought to take the Farnese Atlas seriously as a source of ancient data."

The sculpture offers hints that it may have serious scientific underpinnings. The globe has horizontal lines for the Equator, the Tropics of Cancer and Capricorn, and the Arctic and Antarctic Circles. An angled circle marks the ecliptic, the plane defined by the orbits of the planets, and vertical lines mark colures, the equivalent of lines of longitude on a map of the Earth.

Some historians had speculated that the sculptor might have consulted the work of Ptolemy, who lived about 250 years after Hipparchus, or Aratus, who described the constellations in a poem about a century before Hipparchus. Curiously, no one appears to have suggested Hipparchus' catalog as the reference source.

Dr. Schaefer asked a simple question: in what year were the constellations in the positions depicted on the Farnese Atlas? Because of the wobble in the Earth's axis, the constellations slide around the sky on a 26,000-year cycle.

He noted where parts of constellations that correspond to specific stars intersected with the lines on the globe. For example, the westernmost part of Aries' horn, which corresponds to the star Gamma Ari, touches one of the colures, and the chest of Leo, location of the star Regulus, lies on the Tropic of Cancer line. From photographs he took of the statue during a vacation stop in Naples last year, he calculated the positions of other constellation stars.

"They tell you the time," Dr. Schaefer said. "It's like the big hand moving through the sky."

Using 70 stars, Dr. Schaefer determined that the Farnese Atlas best matches the sky of 125 B.C. give or take 55 years—too early for Ptolemy, too late for Aratus, but almost a perfect match for Hipparchus. In addition, the constellations were placed more accurately than could be deduced from verbal descriptions, Dr. Schaefer said, another argument against Aratus' poem.

He also found that the Farnese Atlas globe agreed in detail with the only surviving work of Hipparchus, a commentary on Aratus and Eudoxes, an earlier astronomer who provided the basis for Aratus' poem. On the other hand, Dr. Schaefer found contradictions between the globe and the writings of Aratus, Eudoxes and Ptolemy.

Dr. Schaefer said Hipparchus made small celestial globes showing the positions of stars in his catalog, and the original Greek sculptor probably consulted one of these globes in chiseling the Farnese Atlas.

"This is a credible demonstration," Dr. Krupp said. "I think this is the first time that a very detailed systematic and careful analysis of this has been performed."

Now that astronomers and science historians seem to have a direct representation of Hipparchus' catalog, they can try to figure out questions like which coordinate system he used. It will also be fodder for a debate as to who was the greater astronomer of antiquity, Hipparchus or Ptolemy?

Dr. Gingerich, who has considered himself in Ptolemy's camp, said, "Now I may have to do some rethinking."

Some Hipparchus supporters even contend that Ptolemy created his star catalog largely by plagiarizing Hipparchus.

Dr. Schaefer diplomatically said that they were both the greatest.

IN REVIEW

1. Explain how the wobble of the Earth's axis can give us clues about observations made by ancient Greek scientists.
2. The Roman statue discussed in the article is a copy of an ancient Greek statue made in the time of Hipparchus. What evidence do scientists have that this copy truly represents the original?
3. What specific lines of evidence did Dr. Schaefer use to pin the statue's drawings to the observations of Hipparchus?
4. What other explanations or arguments could be made that counter the claims of Dr. Schaefer?
5. Discuss the value of studying the contributions by ancient scientists to our understanding of the cosmos.

In recent years there has been an increasing divide between those who trust and those who are suspicious of science. A lack of understanding of the basic characteristics or "hallmarks" of science may be partially to blame. Science requires continual questioning and testing of ideas, and without a fundamental understanding of how observations and models are interweaved to form scientific theories, the process of science may seem uncertain and unreliable. Also, technology, acting as the interface between science and society, while providing many benefits, can have negative consequences. Some people blame science for causing these technological problems.

Does Science Matter?

By William J. Broad and James Glanz
***The New York Times*, November 11, 2003**

Through its rituals of discovery, science has extended life, conquered disease and offered new sexual and commercial freedoms. It has pushed aside demigods and demons and revealed a cosmos more intricate and awesome than anything produced by pure imagination.

But there are new troubles in the peculiar form of paradise that science has created, as well as new questions about whether it has the popular support to meet the future challenges of disease, pollution, security, energy, education, food, water and urban sprawl.

The public seems increasingly intolerant of grand, technical fixes, even while it hungers for new gadgets and drugs. It has also come to fear the potential consequences of unfettered science and technology in areas like genetic engineering, germ warfare, global warming, nuclear power and the proliferation of nuclear arms.

Tension between science and the public has thrown up new barriers to research involving deadly pathogens, stem cells and human cloning. Some of the doubts about science began with the environmental movement of the 1960's.

"The bloom has been coming off the rose since 'Silent Spring,'" said Dr. John H. Gibbons, President Bill Clinton's science adviser, of Rachel Carson's 1962 book on the ravages of DDT. Until then, he said, "People thought of science as a cornucopia of goodies. Now they have to choose between good and bad."

"The urgency," he said, "is to reestablish the fundamental position that science plays in helping devise uses of knowledge to resolve social ills. I hope rationality will triumph. But you can't count on it. As President Chirac said, we've lost the primacy of reason."

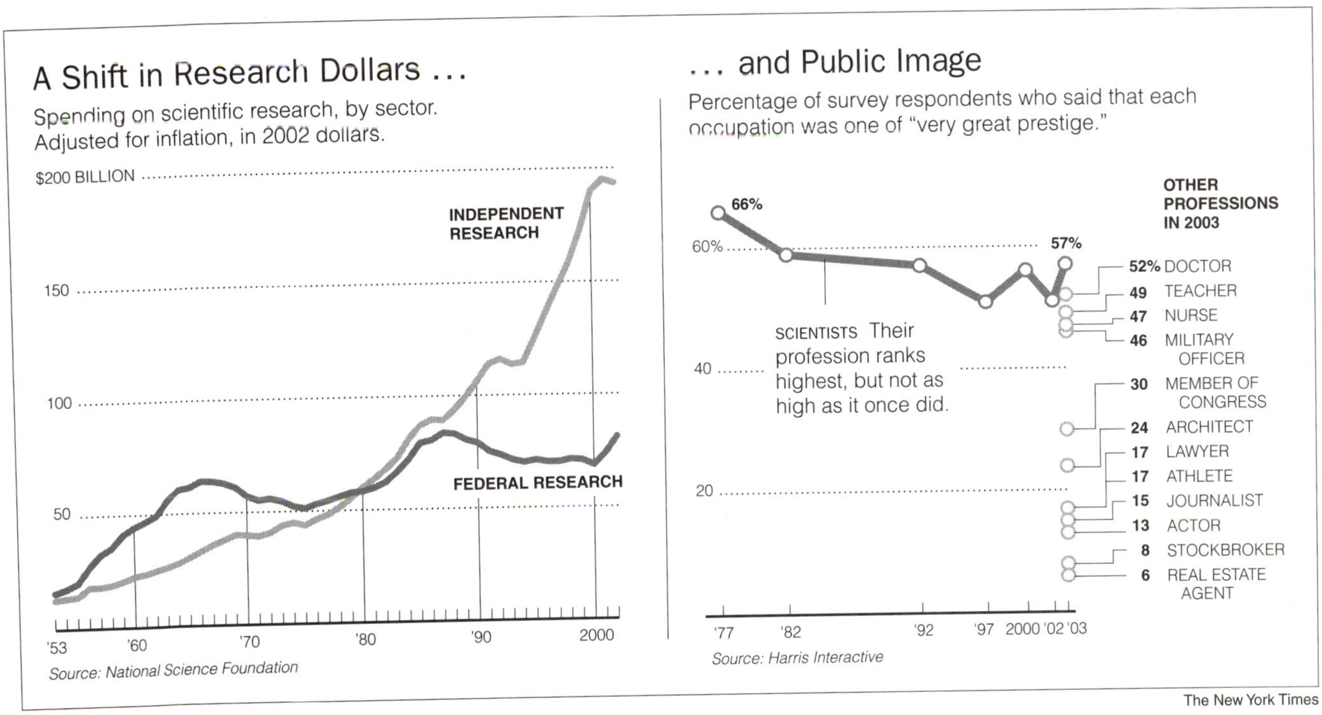

The New York Times

Science has also provoked a deeper unease by disturbing traditional beliefs. Some scientists, stunned by the increasing vigor of fundamentalist religion worldwide, wonder if old certainties have rushed into a sort of vacuum left by the inconclusiveness of science on the big issues of everyday life.

"Isn't it incredible that you have so much fundamentalism, retreating back to so much ignorance?" remarked Dr. George A. Keyworth II, President Ronald Reagan's science adviser.

The disaffection can be gauged in recent opinion surveys. Last month, a Harris poll found that the percentage of Americans who saw scientists as having "very great prestige" had declined nine percentage points in the last quarter-century, down to 57 from 66 percent. Another recent Harris poll found that most Americans believe in miracles, while half believe in ghosts and a third in astrology—hardly an endorsement of scientific rationality.

"There's obviously a kind of national split personality about these things," said Dr. Owen Gingerich, a historian of astronomy at the Harvard-Smithsonian Center for Astrophysics who speaks often of his Christian faith.

"Science gives you very cold comfort at times of death or sickness or so on," Dr. Gingerich said.

In this atmosphere of ambivalence, research priorities have become increasingly politicized, some scientists say.

"Right now it's about as bad as I've known it," said Dr. Sidney Drell, a Stanford University physicist who has advised the federal government on national security issues for more than 40 years.

As the world marches into a century born amid fundamentalist strife in oil-producing nations, a divisive political climate in the United States and abroad and ever more sophisticated challenges to scientific credos like Darwin's theory of evolution, it seems warranted to ask a question that runs counter to centuries of Western thought: Does science matter? Do people care about it anymore?

The Context

Breakthroughs And Disenchantment

Clearly, science has mattered a lot, for a long time. Advances in food, public health and medicine helped raise life expectancy in the United States in the past century from roughly 50 to 80 years. So too, world population between 1950 and 1990 more than doubled, now exceeding six billion. Biology discovered the structure of DNA, made test-tube babies and cured diseases. And the decoding of the human genome is leading scientists toward a detailed understanding of how the body works, offering the hope of new treatments for cancer and other diseases.

"For a lot of people, life has gotten better," said Dr. James D. Watson, co-discoverer of the double helix. "You don't know what it was like in 1950. It wasn't just the dreariness of Bing Crosby that made life tough."

In physics, breakthroughs produced digital electronics and subatomic discoveries. American rocket science won the space race, put men on the moon, probed distant planets and lofted hundreds of satellites, including the Hubble Space Telescope.

But major problems also arose: acid rain, environmental toxins, the Bhopal chemical disaster, nuclear waste, global warming, the ozone hole, fears over genetically modified food and the fiery destruction of two space shuttles, not to mention the curse of junk e-mail. Such troubles have helped feed social disenchantment with science.

When the cold war ended, the physical sciences began to lose luster and funding. After spending $2 billion, Congress killed physicists' preeminent endeavor, the Superconducting Super Collider, an enormous particle accelerator.

"Suddenly, Congress wasn't interested in science anymore," said Fred Jerome, a science policy analyst at the New School.

At the same time, industry spending on research soared to twice that of the federal government, about $180 billion last year, according to the National Science Foundation. One result is that Americans see more drugs, cellphones, advanced toys, innovative cars and engineered foods and less news about the fundamental building blocks and great shadowy vistas of the universe.

The main exceptions to the downward trend in the federal science budget are for health and weapons. This year, spending on military research hit $58 billion, higher in fixed dollars than during the cold war.

Meanwhile, other countries are spending more on research, taking some of the glory that America once monopolized. Japan, Taiwan and South Korea now account for more than a quarter of all American industrial patents, according to CHI Research. Europe is working on what will be the world's most powerful atom smasher. The British are now flying the first probe in a quarter century to look for evidence of life on Mars.

The Contradictions

New Challenges, But Also Threats

Despite the explosion in the life sciences, cancer still darkens many lives, and the flowering of biotechnology has fed worries about genetically modified foods and organisms as well as the pending reinvention of what it means to be human. Many people worry that the growing power of genetics will sully the sanctity of human life.

Last month, the President's Council on Bioethics issued a report warning that biotechnology in pursuit of human perfection could lead to unintended and destructive ends. Experts also worry about terrorists using advances in biology for intentional harm, perhaps on vast new scales.

"As this becomes ever easier and cheaper, it's only a matter of time before some misguided people decide to infect the world," said Dr. Philip Kitcher, a philosopher of science at Columbia University. Last month, a panel of the National Academy of Sciences recommended wide review of experiments that could lead to biological weapons.

The physical sciences seem to have lost what was once a good story line. Without the space race and the cold war, and perhaps facing intrinsic limits as well as declining budgets, they are slightly adrift. Some observers worry that physics has entered a phase of diminishing returns. That theme runs through "The End of Science," a 1997 book by John Horgan.

In an interview, Mr. Horgan noted that physicists no longer make nuclear arms and have lost momentum on taming fusion energy, which powers the sun, and on developing a theory of everything, a kind of mathematical glue that would unite the sciences. Abstract physics, he said, "has wandered off into the fantasy land of higher dimensions and superstring theory and has really lost touch with reality."

Other experts disagree, noting that scientific fields rise and fall in cycles and that physics may be poised for new strides. "You can smell discovery in the air," said Dr. Leon M. Lederman, a Nobel laureate in physics and an architect of the supercollider. "The sense of imminent revolution is very strong."

Despite the decline in prestige recorded in the recent Harris poll, scientists still top the list of 22 professions in terms of high status, ahead of doctors, teachers, lawyers and athletes.

"Science is one of the charismatic activities," said Dr. Gerald Holton, a professor of physics and the history of science at Harvard. "This keeps our interest in science at some level even if we are deeply troubled by some aspects of its technical misuse."

Polls by the National Science Foundation perennially identify contradictions. Its latest numbers show that 90 percent of adult Americans say they are very or moderately interested in science discoveries. Even so, only half the survey respondents knew that the Earth takes a year to go around the Sun.

"The easy answer is, 'Oh, I'm interested,'" said Melissa Pollak, a senior analyst at the National Science Foundation. "I'm not quite sure I believe those responses."

The Competition

The Battles Increase Over Darwin's Theory

A simple number jars many scientists: about two-thirds of the public believe that alternatives to Darwin's theory of evolution should be taught in public schools alongside this bedrock concept of biology itself.

The organized opposition to the mainstream theory of evolution has become vastly more sophisticated and influential than it was, say, 25 years ago. The leading foes of Darwin espouse a theory called "intelligent design," which holds that purely random natural processes could never have produced humans. These foes are led by a relatively small group of people with various academic and professional credentials, including some with advanced degrees in science and even university professorships.

Backers of intelligent design say they are simply pointing up shortcomings in Darwin's theory. Scientists have publicly rallied in response, last week staving off an effort at the Texas State Board of Education to have intelligent design taught alongside evolution.

"It just absolutely boggles the mind," said Dr. James Langer, a physicist at the University of California at Santa Barbara who is vice president of the National Academy of Sciences. "I wouldn't want my doctor thinking that intelligent design was an equally plausible hypothesis to evolution any more than I would want my airplane pilot believing in the flat Earth."

Science has, in fact, sold itself from the start as something more than a utilitarian exercise in developing technologies and medicines. Einstein—who often used religious and philosophical language to explain his discoveries—seemed to tell humanity something fundamental about the fabric of existence. More recently, the cosmologist Stephen Hawking said that discovering a better theory of gravitation would be like seeing into "the mind of God."

Such rhetorical flourishes are as much derided as admired by the bulk of working scientists, who as a culture have drifted closer to the thinking of Steven Weinberg, another Nobel Prize winner in particle physics, who famously wrote that "the more the universe seems comprehensible, the more it also seems pointless."

That almost militantly atheistic view helps some observers explain how science has come into bitter conflict with particular religious groups, especially biblical literalists.

"What accentuates the fault line," said Dr. Ernan McMullin, a Roman Catholic priest who is a former director of the history and philosophy of science program at Notre Dame, is that "the scientists see their science being attacked and they immediately rush to the battlements."

"I think they rather enjoy seeing themselves as a persecuted minority instead of as the dominant force in the culture, which they really are," he said.

The Future

Urgent Goals For Governments

Industry looks to short-term goals and has proven highly adept at using science to take care of itself and consumers. A far more uncertain issue is whether the federal government can successfully address issues of human welfare that lie well beyond the industrial horizon—years, decades and even centuries ahead.

"Science is still the wellspring of new options," Dr. Gibbons said. "How else are we going to face the issues of the 21st century on things like the environment, health, security, food and energy?"

Some experts believe that despite the gnawing doubts today, the world will be ever more inclined to seek scientific answers to those questions in the decades to come. "It will probably accelerate," said Dr. John H. Marburger III, President Bush's science adviser, "because it will become increasingly obvious that we need this steady infusion of results to sustain our ability to cope with all these social problems."

An urgent goal, experts say, is to develop new sources of energy, which will become vitally important as oil becomes increasingly scarce. Another is to better understand the nuances of climate change, for instance, how the sun and ocean affect the atmosphere. Such work is in its infancy. Another is to develop ways of countering the spread of nuclear arms and germ weapons.

The world will also need a new science of cities, to help coordinate planning in areas like waste, water use, congestion, highways, hazard mitigation and pollution control.

"It's going to take a lot of work," said Dr. Grant Heiken, an editor of "Earth Science in the City," a collection of essays just published by the American Geophysical Union in Washington. The number of urban dwellers is expected to grow from three billion now to five billion by 2025.

"I don't know if we'll get a new science," Dr. Heiken said, "but we damn well better."

Dr. Richard E. Smalley, a Rice University professor and Nobel laureate in chemistry, argues that new technologies and conservation can probably solve the world's energy needs. But success, he said, requires a new army of scientists and engineers.

Like others, Dr. Smalley worries about a significant shift in the demographics of American graduate schools in science and engineering. By 1999, according to the latest figures from the National Science Foundation, the number of foreign students in full-time engineering programs had soared so high that it exceeded, for the first time, the steeply declining number of Americans.

"Where the bright kids and the big action are is in Asia," Dr. Smalley said. "That's great for them. It is not what I would hope for our country and our economic well-being or our national security."

Whether the complex challenges of today generate a new era of scientific greatness, several scientists said, may depend on how a deeply conflicted public answers the question of whether science still matters.

In many ways, it all boils down to "a schism between people who have accepted the modern scientific view of the world and the people who are fighting that," said Dr. David Baltimore, the Nobel Prize-winning biologist who is president of the California Institute of Technology.

"Scientists are presenting a much more complicated, much less ethically grounded view of the world, and it's hard for people to take that in," he added.

Some experts warn that if support for science falters and if the American public loses interest in it, such apathy may foster an age in which scientific elites ignore the public weal and global imperatives for their own narrow interests, producing something like a dictatorship of the lab coats.

"For any man to abdicate an interest in science," Jacob Bronowski, the science historian, wrote, "is to walk with open eyes towards slavery."

IN REVIEW

1. List some of the causes of the tension between science and the public.
2. What are some ways that religion and science have clashed in the past? How are these clashes different from today's?
3. How do political leaders use their attitudes about science to shape political agendas?
4. List some of the ways that scientific breakthroughs have benefited the public and some ways that such breakthroughs have caused significant problems. In your opinion, have the benefits outweighed the problems, or vice versa?
5. How much should the government spend to support scientific research? Should the government spend more if competing foreign governments are also spending more?

In the 1,000 years between the fall of the Roman Empire and the European Renaissance, much scientific thought was maintained and advanced by Islamic scholars. Indeed, without these Arabic and Persian scientists, the ideas of the Ancient Greeks might well have perished. In particular, Middle Eastern astronomers developed new instruments for observing the night sky and were responsible for naming many of the stars visible by the unaided human eye. Their observations and scientific theories also laid the foundation for the Copernican revolution and the beginnings of modern astronomy and physics.

How Islam Won, and Lost, the Lead in Science

By Dennis Overbye
***The New York Times,* October 30, 2001**

Nasir al-Din al-Tusi was still a young man when the Assassins made him an offer he couldn't refuse.

His hometown had been devastated by Mongol armies, and so, early in the 13th century, al-Tusi, a promising astronomer and philosopher, came to dwell in the legendary fortress city of Alamut in the mountains of northern Persia.

He lived among a heretical and secretive sect of Shiite Muslims, whose members practiced political murder as a tactic and were dubbed hashishinn, legend has it, because of their use of hashish.

Although al-Tusi later said he had been held in Alamut against his will, the library there was renowned for its excellence, and al-Tusi thrived there, publishing works on astronomy, ethics, mathematics and philosophy that marked him as one of the great intellectuals of his age.

But when the armies of Halagu, the grandson of Genghis Khan, massed outside the city in 1256, al-Tusi had little trouble deciding where his loyalties lay. He joined Halagu and accompanied him to Baghdad, which fell in 1258. The grateful Halagu built him an observatory at Maragha, in what is now northwestern Iran.

Al-Tusi's deftness and ideological flexibility in pursuit of the resources to do science paid off. The road to modern astronomy, scholars say, leads through the work that he and his followers performed at Maragha and Alamut in the 13th and 14th centuries. It is a road that winds from Athens to Alexandria, Baghdad, Damascus and Córdoba, through the palaces of caliphs and the basement laboratories of alchemists, and it was traveled not just by astronomy but by all science.

Commanded by the Koran to seek knowledge and read nature for signs of the Creator, and inspired by a treasure trove of ancient Greek learning, Muslims created a society that in the Middle Ages was the scientific center of the world. The Arabic language was synonymous with learning and science for 500 hundred years, a golden age that can count among its credits the precursors to modern universities, algebra, the names of the stars and even the notion of science as an empirical inquiry.

"Nothing in Europe could hold a candle to what was going on in the Islamic world until about 1600," said Dr. Jamil Ragep, a professor of the history of science at the University of Oklahoma.

It was the infusion of this knowledge into Western Europe, historians say, that fueled the Renaissance and the scientific revolution.

"Civilizations don't just clash," said Dr. Abdelhamid Sabra, a retired professor of the history of Arabic science who taught at Harvard. "They can learn from each other. Islam is a good example of that." The intellectual meeting of Arabia and Greece was one of the greatest events in history, he said. "Its scale and consequences are enormous, not just for Islam but for Europe and the world."

The observatory at Maragha, built for al-Tusi.

But historians say they still know very little about this golden age. Few of the major scientific works from that era have been translated from Arabic, and thousands of manuscripts have never even been read by modern scholars. Dr. Sabra characterizes the history of Islamic science as a field that "hasn't even begun yet."

Islam's rich intellectual history, scholars are at pains and seem saddened and embarrassed to point out, belies the image cast by recent world events. Traditionally, Islam has encouraged science and learning. "There

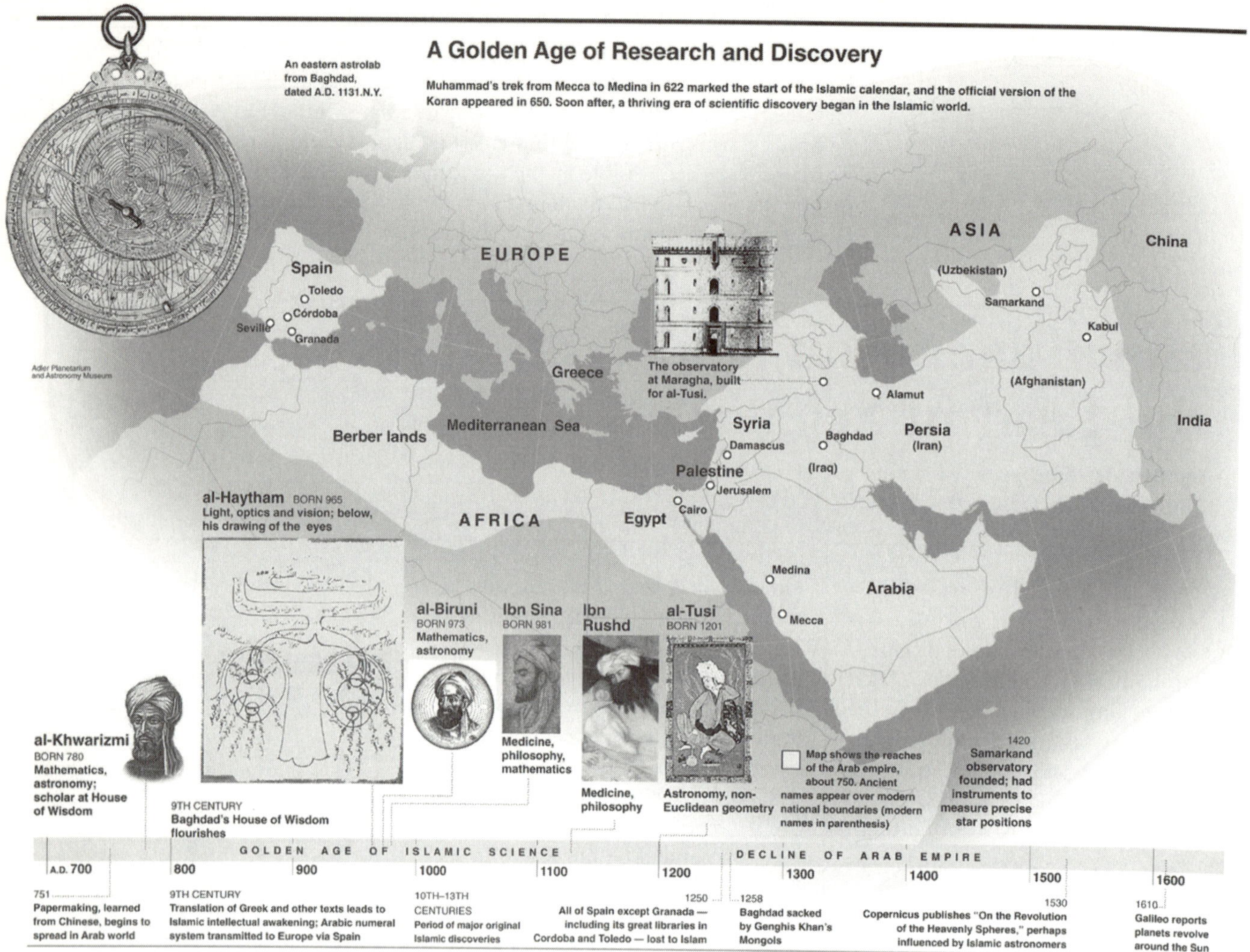

Sources: The Columbia History of the World, History of Islamic Science

The New York Times

is no conflict between Islam and science," said Dr. Osman Bakar of the Center for Muslim-Christian Understanding at Georgetown.

"Knowledge is part of the creed," added Dr. Farouk El-Baz, a geologist at Boston University, who was science adviser to President Anwar el-Sadat of Egypt. "When you know more, you see more evidence of God."

So the notion that modern Islamic science is now considered "abysmal," as Abdus Salam, the first Muslim to win a Nobel Prize in Physics, once put it, haunts Eastern scholars. "Muslims have a kind of nostalgia for the past, when they could contend that they were the dominant cultivators of science," Dr. Bakar said. The relation between science and religion has generated much debate in the Islamic world, he and other scholars said. Some scientists and historians call for an "Islamic science" informed by spiritual values they say Western science ignores, but others argue that a religious conservatism in the East has dampened the skeptical spirit necessary for good science.

The Golden Age

When Muhammad's armies swept out from the Arabian peninsula in the seventh and eighth centuries, annexing territory from Spain to Persia, they also annexed the works of Plato, Aristotle, Democritus, Pythagoras, Archimedes, Hippocrates and other Greek thinkers.

Hellenistic culture had been spread eastward by the armies of Alexander the Great and by religious minorities, including various Christian sects, according to Dr. David Lindberg, a medieval science historian at the University of Wisconsin.

The largely illiterate Muslim conquerors turned to the local intelligentsia to help them govern, Dr. Lindberg said. In the process, he said, they absorbed Greek learning that had yet to be transmitted to the West in a serious way, or even translated into Latin. "The West had a thin version of Greek knowledge," Dr. Lindberg said. "The East had it all."

In ninth-century Baghdad the Caliph Abu al-Abbas al-Mamun set up an institute, the House of Wisdom, to translate manuscripts. Among the first works rendered into Arabic was the Alexandrian astronomer Ptolemy's "Great Work," which described a universe in which the Sun, Moon, planets and stars revolved around Earth; Al-Magest, as the work was known to Arabic scholars, became the basis for cosmology for the next 500 years.

Jews, Christians and Muslims all participated in this flowering of science, art, medicine and philosophy, which endured for at least 500 years and spread from Spain to Persia. Its height, historians say, was in the 10th and 11th centuries when three great thinkers strode the East: Abu Ali al-Hasan ibn al-Haytham, also known as Alhazen; Abu Rayham Muhammad al-Biruni; and Abu Ali al-Hussein Ibn Sina, also known as Avicenna.

Al-Haytham, born in Iraq in 965, experimented with light and vision, laying the foundation for modern optics and for the notion that science should be based on experiment as well as on philosophical arguments. "He ranks with Archimedes, Kepler and Newton as a great mathematical scientist," said Dr. Lindberg.

The mathematician, astronomer and geographer al-Biruni, born in what is now part of Uzbekistan in 973, wrote some 146 works totaling 13,000 pages, including a vast sociological and geographical study of India.

Ibn Sina was a physician and philosopher born near Bukhara (now in Uzbekistan) in 981. He compiled a million-word medical encyclopedia, the Canons of Medicine, that was used as a textbook in parts of the West until the 17th century.

Scholars say science found such favor in medieval Islam for several reasons. Part of the allure was mystical; it was another way to experience the unity of creation that was the central message of Islam.

"Anyone who studies anatomy will increase his faith in the omnipotence and oneness of God the Almighty," goes a saying often attributed to Abul-Walid Muhammad Ibn Rushd, also known as Averroes, a 13th-century anatomist and philosopher.

Knocking on Heaven's Door

Another reason is that Islam is one of the few religions in human history in which scientific procedures are necessary for religious ritual, Dr. David King, a historian of science at Johann Wolfgang Goethe University in Frankfurt, pointed out in his book "Astronomy in the Service of Islam," published in 1993. Arabs had always been knowledgeable about the stars and used them to navigate the desert, but Islam raised the stakes for astronomy.

The requirement that Muslims face in the direction of Mecca when they pray, for example, required knowledge of the size and shape of the Earth. The best astronomical minds of the Muslim world tackled the job of producing tables or diagrams by which the qibla, or sacred directions, could be found from any point in the Islamic world. Their efforts rose to a precision far beyond the needs of the peasants who would use them, noted Dr. King.

Astronomers at the Samarkand observatory, which was founded about 1420 by the ruler Ulugh Beg, measured star positions to a fraction of a degree, said Dr. El-Baz.

Islamic astronomy reached its zenith, at least from the Western perspective, in the 13th and 14th centuries, when al-Tusi and his successors pushed against the limits of the Ptolemaic world view that had ruled for a millennium.

According to the philosophers, celestial bodies were supposed to move in circles at uniform speeds. But the beauty of Ptolemy's attempt to explain the very ununiform motions of planets and the Sun as seen from Earth was marred by corrections like orbits within orbits, known as epicycles, and geometrical modifications.

Al-Tusi found a way to restore most of the symmetry to Ptolemy's model by adding pairs of cleverly designed epicycles to each orbit. Following in al-Tusi's footsteps, the 14th-century astronomer Ala al-Din Abul-Hasan ibn al-Shatir had managed to go further and construct a completely symmetrical model.

Copernicus, who overturned the Ptolemaic universe in 1530 by proposing that the planets revolved around the Sun, expressed ideas similar to the Muslim astronomers in his early writings. This has led some historians to suggest that there is a previously unknown link between Copernicus and the Islamic astronomers, even though neither ibn al-Shatir's nor al-Tusi's work is known to have ever been translated into Latin, and therefore was presumably unknown in the West.

Dr. Owen Gingerich, an astronomer and historian of astronomy at Harvard, said he believed that Copernicus could have developed the ideas independently, but wrote in Scientific American that the whole idea of criticizing Ptolemy and reforming his model was part of "the climate of opinion inherited by the Latin West from Islam."

The Decline of the East

Despite their awareness of Ptolemy's flaws, Islamic astronomers were a long ways from throwing out his model: dismissing it would have required a philosophical as well as cosmological revolution. "In some ways it was beginning to happen," said Dr. Ragep of the University of Oklahoma. But the East had no need of heliocentric models of the universe, said Dr. King of Frankfurt. All motion being relative, he said, it was irrelevant for the purposes of Muslim rituals whether the sun went around the Earth or vice versa.

From the 10th to the 13th century Europeans, especially in Spain, were translating Arabic works into Hebrew and Latin "as fast as they could," said Dr. King. The result was a rebirth of learning that ultimately transformed Western civilization.

Why didn't Eastern science go forward as well? "Nobody has answered that question satisfactorily," said Dr. Sabra of Harvard. Pressed, historians offer up a constellation of reasons. Among other things, the Islamic empire began to be whittled away in the 13th century by Crusaders from the West and Mongols from the East.

Christians reconquered Spain and its magnificent libraries in Córdoba and Toledo, full of Arab learning. As a result, Islamic centers of learning began to lose touch with one another and with the West, leading to a gradual erosion in two of the main pillars of science—communication and financial support.

In the West, science was able to pay for itself in new technology like the steam engine and to attract financing from industry, but in the East it remained dependent on the patronage and curiosity of sultans and caliphs. Further, the Ottomans, who took over the Arabic lands in the 16th century, were builders and conquerors, not thinkers, said Dr. El-Baz of Boston University, and support waned. "You cannot expect the science to be excellent while the society is not," he said.

Others argue, however, that Islamic science seems to decline only when viewed through Western, secular eyes. "It's possible to live without an industrial revolution if you have enough camels and food," Dr. King said.

"Why did Muslim science decline?" he said. "That's a very Western question. It flourished for a thousand years—no civilization on Earth has flourished that long in that way."

Islamic Science Wars

Humiliating encounters with Western colonial powers in the 19th century produced a hunger for Western science and technology, or at least the economic and military power they could produce, scholars say. Reformers bent on modernizing Eastern educational systems to include Western science could argue that Muslims would only be reclaiming their own, since the West had inherited science from the Islamic world to begin with.

In some ways these efforts have been very successful. "In particular countries the science syllabus is quite modern," said Dr. Bakar of Georgetown, citing Malaysia, Jordan and Pakistan, in particular. Even in Saudi Arabia, one of the most conservative Muslim states, science classes are conducted in English, Dr. Sabra said.

Nevertheless, science still lags in the Muslim world, according to Dr. Pervez Hoodbhoy, a Pakistani physicist and professor at Quaid-e-Azam University in Islamabad, who has written on Islam and science. According to his own informal survey, included in his 1991 book "Islam and Science, Religious Orthodoxy and the Battle for Rationality," Muslims are seriously underrepresented in science, accounting for fewer than 1 percent of the world's scientists while they account for almost a fifth of the world's population. Israel, he reports, has almost twice as many scientists as the Muslim countries put together.

Among other sociological and economic factors, like the lack of a middle class, Dr. Hoodbhoy attributes the malaise of Muslim science to an increasing emphasis over the last millennium on rote learning based on the Koran.

"The notion that all knowledge is in the Great Text is a great disincentive to learning," he said. "It's destructive if we want to create a thinking person, someone who can analyze, question and create." Dr. Bruno Guideroni, a Muslim who is an astrophysicist at the National Center for Scientific Research in Paris, said, "The fundamentalists criticize science simply because it is Western."

Other scholars said the attitude of conservative Muslims to science was not so much hostile as schizophrenic, wanting its benefits but not its world view. "They may use modern technology, but they don't deal with issues of religion and science." said Dr. Bakar.

One response to the invasion of Western science, said the scientists, has been an effort to "Islamicize" science by portraying the Koran as a source of scientific knowledge.

Dr. Hoodbhoy said such groups had criticized the concept of cause and effect. Educational guidelines once issued by the Institute for Policy Studies in Pakistan, for example, included the recommendation that physical effects not be related to causes.

For example, it was not Islamic to say that combining hydrogen and oxygen makes water. "You were supposed to say," Dr. Hoodbhoy recounted, "that when you bring hydrogen and oxygen together then by the will of Allah water was created."

Even Muslims who reject fundamentalism, however, have expressed doubts about the desirability of following the Western style of science, saying that it subverts traditional spiritual values and promotes materialism and alienation.

"No science is created in a vacuum," said Dr. Seyyed Hossein Nasr, a science historian, author, philosopher and professor of Islamic studies at George Washington University, during a speech at the Massachusetts Institute of Technology a few years ago. "Science arose under particular circumstances in the West with certain philosophical presumptions about the nature of reality."

Dr. Muzaffar Iqbal, a chemist and the president and founder of the Center for Islam and Science in Alberta, Canada, explained: "Modern science doesn't claim to address the purpose of life; that is outside the domain. In the Islamic world, purpose is integral, part of that life."

Most working scientists tend to scoff at the notion that science can be divided into ethnic, religious or any other kind of flavor. There is only one universe. The process of asking and answering questions about nature, they say, eventually erases the particular circumstances from which those questions arise.

In his book, Dr. Hoodbhoy recounts how Dr. Salam, Dr. Steven Weinberg, now at the University of Texas, and Dr. Sheldon Glashow at Harvard, shared the Nobel Prize for showing that electromagnetism and the so-called weak nuclear force are different manifestations of a single force.

Dr. Salam and Dr. Weinberg had devised the same contribution to that theory independently, he wrote, despite the fact that Dr. Weinberg is an atheist while Dr. Salam was a Muslim who prayed regularly and quoted from the Koran. Dr. Salam confirmed the account in his introduction to the book, describing himself as "geographically and ideologically remote" from Dr. Weinberg.

"Science is international," said Dr. El-Baz. "There is no such thing as Islamic science. Science is like building a big building, a pyramid. Each person puts up a block. These blocks have never had a religion. It's irrelevant, the color of the guy who put up the block."

IN REVIEW

1. What are some of the ways that Islam promotes science?
2. List the major advances that Islamic scientists and mathematics made between 500 and 1500 A.D.
3. After Muslim conquerors swept through the Middle East, Northern Africa, and Spain, how did their actions and policies encourage intellectual thought?
4. What are some of the factors that may have prevented Middle Eastern intellectuals from making the advances in science that occurred during the European Renaissance?
5. Why is science fundamentally international, and why does science progress despite differences in religions throughout the world?

Isaac Newton, who first understood and described the basic laws of motion, is from a different time and world from University of Nebraska football. But by looking at how players collide on the field and how the football flies through the air, physicists can illustrate some of the fundamental concepts that also describe the motions of the Moon, planets, and stars. The terms *velocity, acceleration, mass, force,* and *momentum* come alive on the football field in ways appreciated by fans and scientists alike. During games, a physics professor at Nebraska uses football as a "hook" to increase the fans' understanding of Newton's laws of motion and fundamental physics.

Crunch! Oof! Well, That's Physics

By Henry Fountain
***The New York Times,* November 16, 2004**

It's third and long midway through the second quarter, and Baylor's quarterback arcs a pass 30 yards down the field into Nebraska territory. The ball is thrown in front of the intended receiver, however, and two Nebraska defenders converge on it from opposite directions. Their eyes on the ball and not on each other, they collide at nearly full tilt and the ball pops away.

To the 77,000 fans at Memorial Stadium on this October Saturday, all but a handful dressed in red in tribute to their beloved Cornhuskers, this is just a typical bruising hit, made slightly more interesting, and alarming, because it involves two players on their team. But Dr. Timothy Gay sees it differently.

"Wow, cool, a three-body collision!" Dr. Gay said from his seat in the stands. The forces in this encounter are enormous, but the players don't appear to be injured. Their pads and helmets and the shortness of the collision help protect them, and the third "body"—the ball—absorbs a little bit of the momentum.

Weekdays, Dr. Gay is an experimental atomic physicist at the university who spends most of his time smashing electrons in a basement laboratory, studying the way they scatter as a means of understanding what might go on in the plasma of a fusion reactor or a star.

On fall weekends, when the Huskers play, he makes the short walk across campus to Memorial Stadium,

Photo by David Weaver for The New York Times

Dr. Timothy Gay watches college football games through the eyes of a physicist.

to pursue his avocation—football physics.

To watch a football game with Dr. Gay is to view the sport through a different lens, one where talk of fly patterns, blitzes and muffed punts is supplemented by discussions of vector analysis, conservation of momentum and strange forces that can affect the flight of the ball. At Dr. Gay's perch a dozen rows back on the 35-yard line, Isaac Newton is cited as often as Vince Lombardi, and the X's and O's of the game are enhanced by delta-V's and delta-T's.

Back at his lab after the game, he does a quick estimate of the forces involved in that defender-on-defender hit. The players, who weigh about 200 pounds each, are running at about 20 feet per second, and after they collide they bounce back at perhaps half that speed. It's easy to calculate the acceleration—change in velocity, delta-V, over change in time, delta-T—and force, which by Newton's second law is proportional to mass times acceleration.

The rough result is that the players encounter a force of about 1,800 pounds and an acceleration of 9 g's, or 9 times the force of gravity. Such forces would be bone-breaking and capillary-draining if applied over time, but in the split-second of this collision the players can withstand them. The third body helps a little too—with its much smaller

Source: Dr. Timothy Gay, University of Nebraska

mass, the football is sent flying toward the sideline by the momentum imparted to it.

Dr. Gay draws a parallel to his work. "The three-body collision—that's the kind of thing I do for a living in the lab," he said. "In atom-molecule collisions, you cannot make a certain chemical reaction go unless you have a third body in there to take up some of the momentum. It's essentially the principle of catalysis."

Dr. Gay is as much a teacher as he is a researcher, and for the past five years has been intent on teaching fans of football something of the science behind it, first with a series of humorous one-minute videos shown on the scoreboard at Nebraska games and now with a book, "Football Physics: The Science of the Game."

"My connection to football is deep because what I do is collisions," he said. "I'm really interested in what happens if I send an electron in, where it's going to scatter to, how much momentum will it transfer."

"You see that all the time in football," he added. "You see guys colliding. Obviously the physics is a bit different. In football we use Newtonian physics, in atomic collisions we use quantum mechanics."

Using Newtonian physics, he explains later in the game why it was so easy for Nebraska to score on a short goal-line plunge.

"It's just the classic advantage of the momentum of the offense," Dr. Gay said. Because the offense knows when the ball is going to be snapped and the defense doesn't, the offensive line has about two-tenths of a second to build up momentum before the defense can react.

In Newtonian terms, momentum is simply mass times speed. The Nebraska linemen are both large (on the order of 300 pounds each) and speedy (they can run a 40-yard dash in about 5 seconds), so in that two-tenths of a second the line has built up a lot of momentum—something like 10,000 pound-mass-feet per second in the sometimes arcane units of physics.

Dr. Gay put it more plainly: "They've got a head of steam so they can just bowl over the defense, which hasn't started moving yet. It's Newton's first law in action."

Dr. Gay traces his interest in football and physics to his years at prep school in Massachusetts. He wasn't good enough to make the squad, but he worked as team manager. At the same time, he took his first serious physics course. The two interests coalesced.

"Plus, this was football in New England in the fall," Dr. Gay said. "It doesn't get any better than that."

Of course, it has gotten better. With degrees from Caltech (where he was good enough, or rather the team was bad enough, that he played on the offensive line) and the University of Chicago, Dr. Gay has for the past 11 years been at Nebraska, home to one of college football's blue-chip programs and some of its most loyal fans, who have sold out every home game since 1962. Lincoln is the state capital and a college town, but every fall it is swept up in football mania. Even the portable toilets are done up in Cornhusker red.

This year, after decades of a "three yards and a cloud of dust" philosophy, in which a strong running game is crucial, Nebraska has a new coach and a new way of doing things. It's called the West Coast offense, and it relies far more on the passing game, with the receivers running quick, precise routes and the quarterback timing his throws.

In physics terms, the team has gone from relying on mass and force to emphasizing kinematics, the science of motion and time. The problem is, most of the players were recruited for the old system. So in football terms, the team has gone from good to mediocre—at least by Nebraska's standards.

In the previous week's game, momentum may have been conserved, but pride was not. "Everybody knew we were going to have a tough year," Dr. Gay said. "But we didn't expect to lose, 70-10, to Texas Tech."

This week, though, Nebraska is having an easier time with Baylor, a perennially weak opponent. After a game in which fans pleaded with the team not to throw the football, Nebraska has resorted to more of a ground attack, led initially by running back Cory Ross, who at 5 feet 6 inches and 195 pounds has a low center of mass and is hard to bring down. The Cornhusker line is also throwing its mass around, and Baylor is getting the worst of most collisions. "They're losing the battle of Newton's first law," Dr. Gay said.

That first law—and the second and third, for that matter—are old hat to physicists. "Many of my colleagues say, 'Tim, why are you doing this? This really isn't a very interesting problem, this football physics,'" Dr. Gay said.

"In a narrow sense, they're right," he added. "It's not like we're discovering new physics."

But while it's often described as a collision sport, football is not just about hitting and getting hit. The flight of the ball, for one thing, presents interesting issues. "In this case, I really think there is something new to be learned," he said.

On a basic level, a ball's trajectory has much to do with how tight the spiral is. A wobbly pass presents more surface to the wind, incurring more drag and failing to travel as far. That's one reason, no doubt, why that second-quarter Baylor pass fell short of its target. A tight spiral generally requires that the ball be thrown faster, for reasons of torque. "The harder you throw it, the more torque you apply as it leaves your hand, so it spins faster," Dr. Gay said. "That means that it's more stabilized, and you get a tighter spiral."

Dr. Gay was particularly interested in something an old prep school colleague, now a coach for the New England Patriots, told him about punts—how they tend to drift to the right or the left, depending on the direction of spin and whether they "turn over," dipping front-end down after they reach the high point of their arc.

Drawing on work by another physicist, Dr. Marianne Breinig of the University of Tennessee, Dr. Gay described the forces at work. As the spinning ball moves through the air, one side is moving in the same direction as the air moving past it, while the other is moving in the opposite direction. This results in different relative velocities, and "you actually get a pileup of air on one side," he said.

"Because the friction and drag force are bigger, you get turbulence and a force that shoves the ball." This is the Magnus force, which is what makes a spinning baseball curve.

The Magnus effect doesn't create problems on kickoffs and field goal attempts, he said, because the ball is rotating on a different axis, end over end. It might cause the ball to move up or down, but not side to side, so accuracy wouldn't be affected.

As physics concepts go, the Magnus force is fairly ethereal. Back on the field, the situation is much more concrete: Nebraska is thumping Baylor. The Huskers go on to win by 59-27, aided by Ross and another back, Brandon Jackson, who have scored three touchdowns between them. Dr. Gay, objective physicist, is also a subjective, and happy, fan, left to wonder what it is that makes players like Ross and Jackson so good.

Science, he finds, offers only a partial explanation. "As biomechanical machines they're so complex," he said. "You can talk about overall patterns and features and stuff like that, but it's difficult to say why one running back is better than another.

"You know, heart, the will to win. Physics isn't going to touch that."

IN REVIEW

1. How are Newton's laws of motion demonstrated by the football game between Nebraska and Baylor?
2. The collisions among the football players result in transfers of momentum. Explain how momentum is conserved in these collisions.
3. In your opinion, should the physicist use football to illustrate physics to the fans? Explain.
4. Summarize how Newton's laws of motion explain why Nebraska beat Baylor in the football game.
5. How do frictional and drag forces affect the flight of the football? How are Newton's first and third laws demonstrated by the Magnus effect, which affects thrown, but not kicked, footballs during flight?

Before Albert Einstein stole the spotlight, Isaac Newton was the undisputed "King of Physics." Newton's laws of motion and universal gravitation gave scientists the fundamental equations to understand the workings of the solar system, and ultimately enabled engineers to design the spacecraft that enabled humans to land safely on the Moon and robots to land on Venus, Mars, and Saturn's moon Titan. Behind the genius of Newton is a compelling story of a man who forever changed our view of the universe—a story that ultimately was captured in literature, art, poetry, and music. In this way, the influence of Newton advanced not only science, but cultural history as well.

The Man Who Grasped The Heavens' Gravitas

By John Noble Wilford
***The New York Times,* October 8, 2004**

Correction Appended

If Einstein is today's personification of scientific genius, he inherited that exalted role from none other than Isaac Newton, of whom it was said that this was "the greatest and the luckiest of mortals."

In the tribute, credited to the 18th-century French mathematician Joseph-Louis Lagrange, Newton (1642–1727) was deemed the greatest because he discovered the law of universal gravitation and the luckiest because there was only one universe. His brilliance extended not only to the motions of worlds and falling apples, but to an early system of the calculus and a radical new theory of light and color.

Newton's stature as one of the greatest figures in the history of science, and the influence of his ideas on the wider culture for more than two centuries, is the subject of a thoughtful and engaging exhibition at the New York Public Library. The show, "The Newtonian Moment: Science and the Making of Modern Culture," opens today and will run through Feb. 5.

"Newton was immensely curious and obviously immensely talented," said Dr. Mordechai Feingold, a history professor at the California Institute of Technology who is curator of the exhibition, as he conducted a tour of the hundreds of rare books, prints, instruments and other displays of Newtonian science.

A number of Newton's manuscripts from the Cambridge University Library, including a first edition of his most famous work, "Principia," bearing the author's corrections and additions for the next printing, have never before been shown in the United States. Dr. Feingold has also written a companion book to the exhibition, to be published this fall by Oxford University Press.

The Newtonian Moment began in the plague years of 1665-66 when the young scholar first formulated many of his revolutionary ideas of mathematics, optics and mechanics. He had already absorbed the essential science of Descartes, Galileo and Robert Hooke. As Dr. Feingold observed, much of Newton's genius consisted of "his remarkable ability to simultaneously consume and transform any knowledge he acquired."

Newton himself attributed at least some of his success to the fact that he had stood "on the shoulders of giants."

Newton came to prominence in 1671 when a small telescope he designed and built won him election to the Royal Society. Then Edmond Halley, he of comet fame, went to Cambridge to ask Newton's advice on the shape of orbits traveled by planets. Ellipses, Newton replied forthwith. He had already figured it out.

Impressed, Halley persuaded a reluctant Newton to write what became his magnum opus, "Principia," published in 1687. His other great work, "Opticks," followed in 1704, describing how a beam of light, when passed through a prism, dispersed into the many colors of the visual spectrum.

The exhibition dwells at length on the international response to Newton's science. Germans contended that Leibniz was the true inventor of calculus, and his principles are indeed the basis for today's calculus. The French stoutly defended their Descartes, whose universe was full of matter whirling in vortices. The ascendant Dutch universities served as arbiters in Newton's ultimate triumph.

The French, to their credit, eventually organized expeditions to Lapland and Peru that confirmed a prediction of Newtonian gravity: the Earth is an oblate spheroid, flattened at the poles and bulging at the equator.

The complex character of Newton is also explored in the exhibition. He was a pious man who dabbled in chronologies of biblical history, though he apparently held unorthodox views on Christian doctrine like the Trinity. He also devoted much effort to alchemy, the practice of trying to turn base metals to gold, which makes Newton a central character in a new novel, "The System of the World," by Neal Stephenson (William Morrow). Dr. Feingold suggests that Newton's alchemy was probably the contemporary name for what today would be standard chemistry.

In many ways, the exhibition's most fascinating displays highlight the dissemination of Newtonian ideas beyond the narrow world of science, which contributed to their transforming influence on the entire culture and the idolization of genius himself. He came to be considered, in Dr. Feingold's phrase, "the acme of human possibility."

"Science was a minor partner in culture in 1600," Dr. Feingold said on the tour. "By 1800, it was the major part of culture."

For more than a century after his death in 1727, Newton was immortalized by poets, sculptors, painters and other worthies whose works are on display. Alexander Pope's familiar couplet eulogized Newton: "Nature and Nature's Laws lay hid at Night. / God said, Let Newton be! and All was Light." He is similarly memorialized in prints and reproductions by William Hogarth and numerous other artists.

Newton's bust had a prominent place in aristocratic gardens, and often appeared in the background of portraits of scientists and men of letters, notably one of Benjamin Franklin. Even Newton's critics recognized his towering intellect. Under a color print by William Blake, showing Newton seated on a rock in the "sea of time and space," the description label notes that for Blake, "Newton is the misguided genius whose mechanical universe left no room for the imagination or for God, but who could ultimately find a prominent place in heaven."

Another Newton legacy, amply illustrated in the displays, was the popularization of science. The trend became especially widespread in the 1740's with the fascination about the nature of electricity. A print of one electrical experiment anticipated Rube Goldberg. A boy, suspended by silk cords, is shown drawing an electric charge from a generator. He transmits the charge to the girl standing on dried pitch, allowing her to attract and repel chaff.

Several arresting displays show the actual devices or drawings of mechanical models of the solar system that were popular in the post-Newton period. Such an elaborate device, known as the orrery, illustrated the relative size and motion of heavenly bodies in accordance with Newton's laws of gravity.

Writers like Francesco Algarotti also facilitated the passage of Newton's ideas to the general public. He translated the recondite language of Newton's "Principia" into a dialogue format interspersed with amusing digressions. The book was written specifically to appeal to women. Other books of the time, like the Tom Telescope series, introducted Newton to children.

The most famous popularizer of Newtonian science was Voltaire, who in 1738 published the book "Elémens de la Philosophie de Neuton." The book's allegorical frontispiece typifies the virtual deification of Newton. Voltaire is seen writing at his desk. Over him a shaft of light from heaven, the light of truth, passes through Newton to the author's lover and collaborator, Madame du Ch,telet. She reflects the light onto the inspired Voltaire.

The acknowledgment of Madame du Châtelet's role ("Minerva dictated and I wrote," Voltaire said) was well deserved. Voltaire's muse happened to be a versatile natural philospher who translated "Principia" into French, accompanied by an extensive commentary, and made no attempt to conceal her talents. She once told Frederick the Great of Prussia, "It may be that there are metaphysicians and philosophers whose learning is greater than mine, although I have not met them."

Voltaire's book was a great success. He commanded a wide audience, Dr. Feingold said, because "he was neither a mathematician nor a physicist, but a literary giant aloof from the academic disputes over Newtonian ideas." In other words, the historian added, Voltaire's amateurism in science "was a source of his contemporary appeal, demonstrating for the first time the accessibility of Newton's ideas to nonspecialists."

So a journalist who makes his living reporting and explaining the labors of scientists comes away from the exhibition both elevated and humbled. He can count Voltaire among his most celebrated forerunners. Even science journalists stand on the shoulders of giants.

He Saw the Light

"The Newtonian Moment: Science and the Making of Modern Culture," is on view through Feb. 5 in the Gottesman Exhibition Hall in the Humanities and Social Sciences Library of the New York Public Library, Fifth Avenue and 42nd Street. Hours: Tuesdays and Wednesdays, 11 a.m. to 7:30 p.m.; Thursdays through Saturdays, 10 a.m. to 6 p.m. Admission is free. Information: (212) 869-8089

Two lectures accompany the exhibition; tickets to each are $10. "Man of the Moment: How Newton Moved Mathematics to the Top of the Scientific Agenda," by Lisa Jardine, a professor of Renaissance Studies at Queen Mary, University of London will be given Nov. 30 at 6:30 p.m. "The Scientist as Scholar: Newton as Historian," by Anthony Grafton, a professor of history at Princeton University, will be on Jan. 5 at 6:30 p.m.

Correction: October 18, 2004, Monday. An article in Weekend on Oct. 8 about the Isaac Newton exhibition at the New York Public Library included the library's misstatement of part of a couplet by Alexander Pope. His eulogy of Newton read "Nature and Nature's Laws lay hid in Night. / God said, Let Newton be! and all was Light." (Pope did not write "hid at Night.")

IN REVIEW

1. Newton attributes some of his genius to the earlier work of others. How is this statement a reflection of the processes of science?
2. What life events described in the article are attributable to some of Newton's scientific breakthroughs? In your opinion, would he have made his important discoveries about gravitation, force, and motion without these life events? Explain.
3. Newton was immortalized by "poets, sculptors, and painters" after his death. Explain why someone would want to use art to immortalize the works of a scientist.
4. Newton was responsible, in large part, for the popularization of science in the 1700s and 1800s. How did artists using Newton's scientific theories help in this popularization?
5. Einstein eclipsed Newton in popularity in the early 1900s. What aspects of Einstein's work might have made him more popular? What societal and technological changes in the early 1900s might also have contributed to Einstein's popularity?

Gravity is the force of attraction between masses. Earth, with its tremendous mass, creates a force of gravity that is responsible both for keeping us safely on the planet's surface and for keeping the Moon in orbit. However, the Earth's relatively large gravitational force makes it difficult, and therefore costly, to launch spacecraft into orbit and beyond. Indeed, some attribute the current delay in human exploration of the solar system to the large costs associated with escaping the Earth's gravity. This idea is clearly illustrated in a quote by science fiction author Robert Heinlein: "Once you get to Earth orbit, you're halfway to anywhere in the solar system." A solution to the high cost of getting objects into Earth's orbit may be the space elevator, which has been promoted by author Arthur C. Clarke for many years and may now be possible because of the creation of certain high-tech materials.

Not Science Fiction: An Elevator to Space

By Kenneth Chang
***The New York Times,* September 23, 2003**

With advances toward ultrastrong fibers, the concept of building an elevator 60,000 miles high to carry cargo into space is moving from the realm of science fiction to the fringes of reality.

This month, the Los Alamos National Laboratory was a sponsor of a conference to ponder the concept. Yet, the keynote address was by a titan of science fiction, Arthur C. Clarke, speaking via satellite from his home in Sri Lanka. "I'm happy that people are taking it more and more seriously," said Mr. Clarke, whose novel "The Fountains of Paradise" (1978) revolved around such a space elevator.

The discovery in 1991 of nanotubes, cylindrical molecules of carbon with many times the strength of steel, turned the idea from a fantastical impossibility to an intriguing possibility that could be realized in as little as a decade or two.

Proponents say the economic and technological advantages of a space elevator over rockets make it inevitable. They predict it will lower the cost of putting a satellite into space from $10,000 a pound to $100.

"As soon as we can build it, we should build it," said Dr. Bryan E. Laubscher, a scientist at Los Alamos who organized the conference. Just as the transcontinental railroad opened the West in the late 1800's, "I feel the space elevator is going to be such a paradigm shift in space access," Dr. Laubscher said.

Photo by Flight Projects Directorate

Scientists and engineers are working on the concept of an elevator to space.

Easier economical access to space might also make practical other grandiose projects like solar power satellites that could collect sunlight and beam energy down to Earth.

The conference, a three-day session here, drew 60 people, a mix of scientists and engineers who are working on the concept, space enthusiasts who wanted to hear more and dilettantes from nearby Los Alamos laboratory attracted by curiosity.

"The first thought is, Is this really going to work?" said Dr. Steven E. Patamia, a researcher at Los Alamos, who was enlisted into performing space elevator calculations a week before the conference. "When you get into it, it begins to make sense. There are a good number of technical issues. They are probably all 'overcomeable.'"

The original idea of a space elevator is more than a century old. In 1895, Konstantin E. Tsiolkovsky, a Russian visionary who devised workable ideas for rocket propulsion and space travel decades before others, proposed a tower thousands of miles high attached to a "celestial castle" in orbit around Earth, with the centrifugal force of the orbiting castle holding up the tower. (Imagine swinging a rope with a rock tied to the end of it.)

But the idea was fundamentally impossible to build. Steel, then the strongest material known, was too heavy and not strong enough to support that weight.

Other scientists periodically revisited and reinvented Tsiolkovsky's idea, inspiring science fiction writers like Mr. Clarke.

Nanotubes spurred NASA to take a more serious look in 1999. A team of scientists envisioned huge cables of nanotubes and magnetically levitated cars traveling up and down. The structure would be so large that it would require grabbing an asteroid and dragging it into Earth orbit to act as the counterweight for holding up the elevator.

To avoid weather, especially lightning, the NASA scientists envisioned the base station as a tower at least 10 miles high.

"We came out of that workshop saying the space elevator is 50 years away," said David V. Smitherman of the Marshall Space Flight Center, who led the study.

Around that time, Dr. Bradley C. Edwards, who was then a scientist at Los Alamos, read an even more pessimistic assessment, that a space elevator would not be built for at least 300 years.

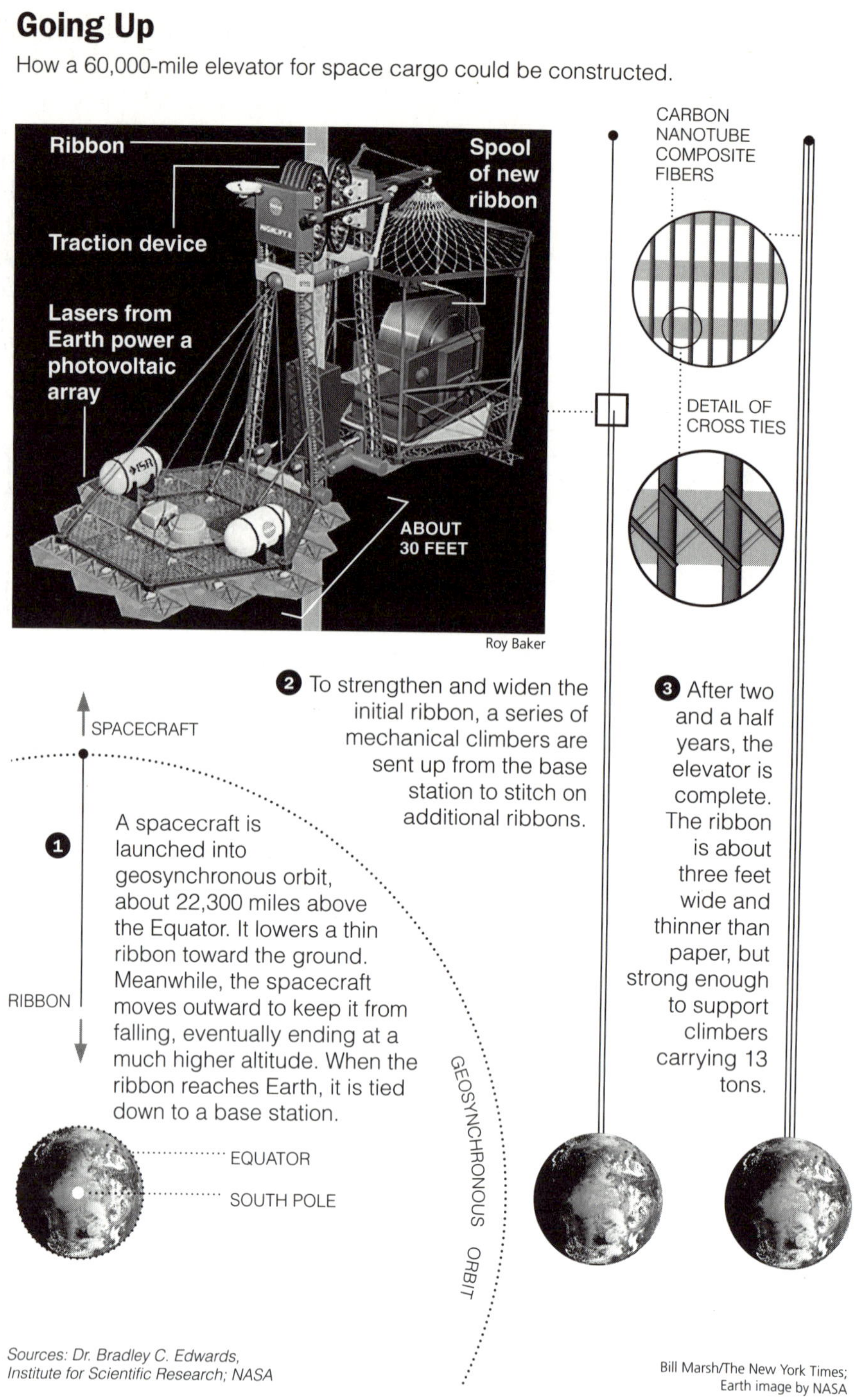

"But there was no information why it couldn't be built," Dr. Edwards said, and he took that as a challenge.

Dr. Edwards simplified the NASA idea to what he calls "the Wright brothers' version," a single ribbon about three feet wide and thinner than a piece of paper, stretching 60,000 miles from Earth's surface.

He sent a proposal to the NASA Institute for Advanced Concepts, which provided him $570,000 to flesh out the ideas. His results are described in a book simply titled "The Space Elevator" (Spageo, 2002).

Instead of using magnetic levitation, the apparatus would lift up to 13 tons of cargo by pulling itself upward with a couple of tanklike treads that squeezed tightly onto the ribbon. Up to eight would ascend the ribbon at any one time, powered by lasers on the ground shining on the solar panels on the rising platforms.

It would take about a week for one to reach geosynchronous orbit, 22,300 miles up, where a satellite circles the Earth in exactly one day, continuously hovering over the same spot on the Earth's surface.

The first elevator would go up only. At the top, the platform would simply be added to the counterweight or be discarded into space.

All the necessary underlying technology exists, Dr. Edwards said, except the material for the ribbon. (The longest nanotube to date is just a few feet long.) But he said he expected that scientists would develop a strong enough nanotube-polymer composite in a few years.

"There's a clear path to building this," said Dr. Edwards, now director of research at the Institute for Scientific Research, an independent organization in Fairmont, W.Va. The institute sponsored the conference with Los Alamos.

Dr. Edwards estimates the cost to build the first elevator at $6.2 billion, although with the uncertainties in forecasting a decade or two of research and development, "doubling this is probably a good first cut."

"It's gotten to the point," he added, "where we can say it's closer to $6 billion than $600 billion."

Building subsequent elevators would be cheaper, $2 billion each, because the first elevator could lift materials.

By comparison, the estimated cost of building and operating the International Space Station is widely expected to exceed $100 billion.

Dr. Edwards says he has reasonable solutions for other concerns. The elevator base could be a movable ocean platform in the eastern equatorial Pacific, hundreds of miles from commercial airline routes and easy to defend from terrorist attacks. Hurricanes never cross the Equator, and lightning is sparse in that region. By moving the base station, the elevator operators could drag the apparatus around low-orbit space debris. An aluminum coating could be added to parts of the ribbon to combat decay from the reaction of oxygen with carbon atoms in the nanotubes.

More futuristically, Dr. Edwards imagines that additional elevators can be built on the Moon or Mars, vastly simplifying and speeding spaceflight through the solar system.

"What a wonderful idea if you could ever make it work," said Gentry Lee, the chief engineer of planetary flight systems at NASA's Jet Propulsion Laboratory in Pasadena, Calif., who pushed for financing Dr. Edwards's initial studies. "It is plausible. It is not implausible. I think that the idea has so much promise a couple of million dollars a year aimed at the enabling technologies is not too much to ask."

At the conference, scientists presented calculations that examined details. Vibrations in the elevator ribbon, which would act like an extremely long plucked guitar string, appeared manageable.

Dr. Anders Jorgensen of Los Alamos raised concerns that as the ribbon swung around through the Earth's magnetic field, it would create strong electric currents. Because of the elevator's relatively slow pace, a larger problem could be that any human passengers would receive dangerous doses of radiation as they passed through pockets of high-energy particles trapped in Earth's magnetic field.

The first space elevator would be built to carry only cargo, not people, Dr. Edwards said. The dangers could be reduced on subsequent elevators by speeding them up or providing shielding through magnetic fields.

The logistics of construction "appear to be an interesting problem, as well," said Carey R. Butler, a program manager at the Institute for Scientific Research. The idea is to launch a spacecraft with the initial spools of ribbon into geosynchronous orbit. As the ribbon unspools and falls to the ground, the spacecraft moves higher to keep the center of mass, or balance point, at the same geosynchronous height. By the time the ribbon reaches the ground, the spacecraft has reached an altitude of 50,000 miles.

After the ribbon is attached to the base station, the spacecraft unfurls another 10,000 miles of ribbon. Then a series of about 230 mechanical construction platforms will ascend to stitch on additional ribbon.

While technologically feasible, years of engineering will still be needed. "There's a lot to be done, obviously," Dr. Edwards said.

Nonetheless, Mr. Clarke, who came up with the idea of using satellites in geosynchronous orbit for communications long before any were launched, thought that he might live to see to live this science fiction idea come true, too.

"I'm 86 now," Mr. Clarke said. "So in 20 years' time, I'll only be 106. So maybe I will see it."

IN REVIEW

1. Describe the basic principles behind the space elevator.
2. How have advances in materials made space elevators possible? What other technological advances are necessary for space elevators to be built?
3. What role have science fiction writers played in promoting the idea of the space elevator? List some other examples of how science fiction helped shape technologies that ultimately became reality.
4. In your opinion, is the cost of building a space elevator justifiable? Explain.
5. Using your knowledge of Newton's laws of motion and universal gravitation, why would a space elevator be a realistic technology for reaching the Moon or Mars?

When seen in the night sky, Mars appears as a red star. Indeed, its crimson color reminded the ancients of blood and, to them, foretold the coming of war. However, when Mars is closest to Earth, its color shifts from red to more of an orange or orange-brown (known as ochre). Cameras on successful U.S. missions that have landed on Mars have also shown Mars in a variety of colors, leaving scientists and the public to ponder what the planet would look like if they were standing on its surface. Part of the problem lies in how the images of Mars are created by the cameras, an issue addressed in a branch of science called Imaging. The proper use of colored filters and computer processing is necessary to represent the color of Mars and other astronomical objects accurately, or at least in a manner that reflects what scientists are trying to communicate with their image.

How the Red Planet Came Down With the Pink Blues

By Kenneth Chang

***The New York Times,* February 10, 2004**

It hardly seemed a surprise that the panoramic photograph of Gusev Crater, the landing site of NASA's Mars rover Spirit, showed a red landscape. After all, Mars is called the red planet because it looks red, a hue imparted by rust in the planet's ubiquitous dust.

But did NASA fiddle with the image to make it look that red? As Mars buffs have pointed out in recent weeks on Web sites like Slashdot.org, a closer look reveals that parts of the rover itself, in the foreground, are oddly garish. Even the color chips placed on the rover to calibrate the color photographs had shifted. What should be bright blue is instead bright pink; what should be bright green is brown.

A few days later, after the rover Spirit made its first roll onto the surface, it took a picture of the empty lander behind it. Again, the Mars buffs zoomed in on a detail. NASA's navy blue logo, often called the "meatball" because of its shape, was now the pinkish hue of rare hamburger. Perhaps more shocking, the spacecraft designers at NASA appeared to have a thing for hot pink: that was the color of the cables strewn around the top of the lander.

What was going on? On Jan. 31, during a lull in the control room at NASA's Jet Propulsion Laboratory, Jason Soderblom, a graduate student at Cornell who is a member of the science team, gave a talk explaining the odd Martian colors.

Photo by NASA/J.P.L./Cornell

The white arrows point to patches of color that changed hue on Mars. The dark patches (top) (bright green and blue in the original photo) on Earth turned lighter (bottom) (brown and pink in the original photo) in Mars photos.

The fundamental challenge in creating color photographs of Mars, he said, is that the cameras on the rovers take only black-and-white pictures, and the art of making color out of black and white never exactly reproduces what the eye sees.

To produce a color photograph, the rover's panoramic camera takes three black-and-white images of a scene, once with a red filter, once with a green filter and once with a blue filter. Each is then tinted with the color of the filter, and the three are combined into a color image.

In assembling the Spirit photographs, however, the scientists used an image taken with an infrared filter, not the red filter. Some blue pigments like the cobalt in the rover color chip also emit this longer-wavelength light, which is not visible to the human eye.

"This is actually true for a lot of blue pigments," Mr. Soderblom said. The infrared is so bright that it washes out the blue. "That turns this thing from a dark blue chip to a vibrant pink," he said. "This is an incredibly bright pink. The same thing is happening with our green color chip."

The infrared-for-red replacement is also "why the NASA meatball is red instead of blue," Mr. Soderblom said. The same is true of the cables. They are actually blue.

For the scientists, there are good reasons to focus on infrared colors rather than the visible red. "Iron dominates mineral color in the visible, and it causes everything to have shades of red," Mr. Soderblom said.

With the infrared filter, the different iron minerals emit different colors, and the camera can better differentiate between them. "We're trying to identify the minerals in the scene, and the way we're doing this is with subtle differences," Mr. Soderblom said.

What the eye sees, even when the colors are right, is not always quite what is happening. Mr. Soderblom showed pictures of the reddish "lily pads" around the landing site of the Spirit's twin, Opportunity, where its air bags bounced. The soil in the bounce marks is actually no more red than the darker surrounding soil, but for reasons not yet known, it emits less blue light. "By being not as bright in the blue, it appears more red to us," he said.

Still, there was no reason for the Spirit to see pink on Mars. When producing the panorama, the camera also used the red filter.

"We just made a mistake," said Dr. James F. Bell III, the lead scientist for the camera. "It's really just a mess-up."

IN REVIEW

1. How do scientists use filters to obtain colored images of Mars?
2. Besides creating images that reflect what the planet might look like, what uses do the filters have in helping scientists analyze the martian surface?
3. How might science-fiction movies influence how scientists and the public view the color of Mars?
4. What clues did those viewing the images have that improper filters might have been used?
5. Looking at colored images of other astronomical objects in the media or in astronomy textbooks, speculate about the methods of manipulation that scientists may have used when constructing the image. In your opinion, should scientists disclose their manipulations in creating the image? Explain.

In February 2004 President Bush announced a new space initiative, directing NASA to redefine its vision to focus on human and robotic exploration of the Moon, Mars, and eventually, bodies farther out in the solar system. Charged with such an initiative, the agency has had to limit its commitment to the Space Shuttle and International Space Station, as well as to science missions predominantly observing objects beyond the solar system. In the name of safety, the Hubble Space Telescope became a victim to NASA's redefined mission. Hubble has been looking into the deepest regions of space for the past 15 years unfettered by distorting effects of the Earth's atmosphere, and the telescope has been due for updated equipment to continue its mission. NASA, however, has decided that future servicing missions to the telescope would put astronauts in a greater degree of danger than its potential future scientific discoveries would warrant.

Panel Urges Shuttle Mission To Help Hubble Telescope

By Warren E. Leary and Dennis Overbye
***The New York Times,* December 9, 2004**

Contradicting NASA policy in a long and bitter debate over science and safety, a panel of the National Academy of Sciences said Wednesday that the agency should send astronauts to repair the Hubble Space Telescope rather than rely on a robotic device.

After six months of study conducted at Congress's request, the committee of 21 experts said that a robotic mission would hold too many uncertainties, that it would probably be ready too late to extend the telescope's life and that it might actually damage the instrument.

As for the risk to a space shuttle crew, the panel said, there is a "very small" difference in safety between a mission to the telescope and a typical mission to the International Space Station. More than two dozen journeys to the station are planned over the next decade.

"A shuttle servicing mission is the best option for extending the life of the Hubble telescope," said Dr. Louis J. Lanzerotti, a physics professor at the New Jersey Institute of Technology and chairman of the panel, assembled by the academy's National Research Council. "It is highly unlikely that the science life of Hubble would be extended with a robotic mission."

There was no immediate response from the National Aeronautics and Space Administration, which touched off an outcry from astronomers and lawmakers last January when its administrator, Sean O'Keefe, announced that he was canceling a shuttle mission to repair the telescope because it would be too dangerous to the crew. Over the summer, Mr. O'Keefe called for the development of a robot that could perform the repairs, without which the telescope is expected to die in orbit in 2007 or 2008.

While the recommendations in the new report are not binding, they will put pressure on NASA to reconsider its position. Representative Sherwood Boehlert, Republican of New York and chairman of the House Science Committee, noted that the findings were "diametrically opposite to those reached by NASA" and said he would hold hearings early next year on the report and all options for the telescope.

Members of the National Academy's panel briefed Mr. O'Keefe and senior NASA managers on their conclusions on Tuesday.

On Wednesday, Robert Mirelson, an agency spokesman, said NASA would need more time to study the report.

"In the meantime, NASA is continuing to examine a variety of methods for extending the Hubble's life," Mr. Mirelson said. The agency will continue to plan for a robotic mission, he said, but will not do anything to preclude a possible shuttle mission.

Earlier in the day, Dr. Lanzerotti, who is also a consultant with Bell Laboratories and Lucent Technologies, said at a news conference that the telescope was one of the greatest achievements of the space program and that keeping it in orbit should be a priority so it could continue making significant discoveries about the universe.

A robotic mission for saving the telescope presents many technical risks because nothing like it has ever been done, requiring the development and testing of new technology.

Astronauts have serviced the telescope four times since it was launched in 1990, and a shuttle mission could be accomplished sooner and with a higher probability of success than one using untried robotic measures, the 135-page report said.

Astronomers said they were delighted by the experts' findings. Dr. Steven Beckwith, director of the Space Telescope Science Institute in Baltimore, which supervises the Hubble's research operations, said of the telescope: "They have affirmed that the future of its science is very bright. It's not an old telescope whose time is gone. If anything, it is becoming more valuable."

Dr. John Bahcall, a physicist at the Institute for Advanced Study in Princeton, N.J., who has been involved with the Hubble program from its beginning, called the new study "a wonderful report" and said of its dismissal of the robotic mission, "Finally, somebody told the king he didn't have any clothes."

Several astronomers said they had long been worried about the effects of a costly robotic mission on other science projects and about the possibility of resulting friction between astronomers and astrophysicists. Estimates for a robotic mission have been as high as $2.6 billion, which would come out of NASA's science budget, perhaps delaying other projects like the study of black holes and dark energy.

Dr. Donald Lamb, a University of Chicago astrophysicist, said, "The tremendous cost of the robotic repair mission would really mean a broad array of other space science missions would be very unlikely to happen."

On Capitol Hill, Senator Barbara A. Mikulski, the Maryland Democrat who helped secure $300 million for NASA's current budget to use for servicing the telescope, said: "NASA has the experience, the technology, and now it has the money. It's time to fix Hubble."

Mr. O'Keefe, the NASA administrator, said last winter that he considered a manned Hubble mission riskier than shuttle flights to the space station. He said a strict interpretation of the findings of the board that investigated the loss of the shuttle Columbia and its crew of seven made him believe that the risk of sending astronauts to the observatory was too high.

But the academy's report notes that NASA is planning 25 to 30 more shuttle missions to the space station and says the risks faced by a crew on a mission to the telescope would be similar to the risk of any single mission to the station.

Roger E. Tetrault, a member of the committee who also served on the Columbia investigation board, said at the news conference that all space travel was risky and that the nation had accepted a certain level of risk with the human program.

"We are saying that going to Hubble as opposed to going to the space station is worth that incremental risk, which is not large," Mr. Tetrault said.

IN REVIEW

1. List the main areas of disagreement between NASA and the panel of experts regarding servicing the Hubble Space Telescope.
2. What are some of the reasons that Congress convened a panel of science experts to look at potential ways to service Hubble?
3. NASA is considering a robotic mission to service Hubble. What are the advantages and disadvantages of repairing Hubble with robots?
4. According to the article, some scientists claim that the nation has accepted a certain level of risk associated with human space exploration. In your opinion, what is an acceptable level of risk for humans in space? Explain.
5. In your opinion, should NASA focus its resources on human space exploration by visiting and colonizing the Moon and Mars, or should the agency focus on purely scientific pursuits such as the Hubble Space Telescope? Explain.

With increasing technological capabilities, scientists have become inundated with data—so much, in fact, that it is increasingly difficult for them to analyze collected data, leaving the possibility that some important scientific truths may remain undiscovered. But scientists are turning back to technology to solve the problem of data overload. Using large databases and Internet software, scientists have created a virtual observatory, which acts as a large repository of information that is easily mined and analyzed to reveal new wonders of the universe. In addition to helping professional scientists, this virtual observatory has opened doors for amateur astronomers, in essence creating a new class of researchers that some call "citizen scientists." No longer an opportunity for only the elite, the virtual observatory and its software tools make science open to all who have a genuine interest.

Telescopes of the World, Unite! A Cosmic Database Emerges

By Bruce Schechter
***The New York Times,* May 20, 2003**

About a year ago a large group of astronomers began to assemble what some of them were calling "the world's best telescope." Their ambitious instrument is still far from complete, but they recently took it for a test run. Within minutes, to their joy and astonishment, they had discovered three or four brown dwarfs, objects that occupy the niche between planet and star.

"It gave me shivers when I heard about it," said Dr. Alex Szalay, a Johns Hopkins astronomer who is one of the telescope's chief architects.

It wasn't the brown dwarfs themselves that excited Dr. Szalay; hundreds of them have been discovered in the past decade. But he and many other astronomers believe that the means used to discover these objects heralds the beginning of a new era of astronomy, and even a new era of science.

Finding a Needle in Two Haystacks

Brown dwarfs are often too dim to be detected, even by powerful telescopes. But when digital sky surveys were combined, new dwarfs quickly emerged from the background.

2MASS | **SDSS**

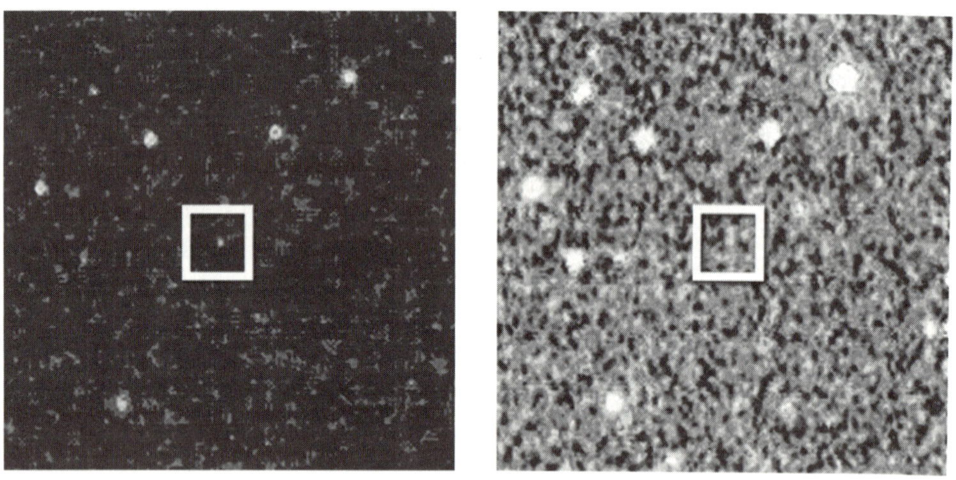

The Two Micron All Sky Survey, or 2MASS, views the sky in the infrared. The object in the center of the square might be a brown dwarf, but since it appears in only one infrared band it might also be "junk."

But the object also appears in the Sloan Digital Sky Survey, which looks at the universe of visible light. The images were compared, and subsequent observations proved that it was indeed a brown dwarf.

Source: National Virtual Observatory

The New York Times

The telescope that Dr. Szalay and his colleagues have constructed is not built of glass and metal. It is a virtual observatory, consisting of terabytes of data collected by dozens of telescopes on Earth and in space, and the software necessary to mine these data for scientific gems.

Like much of the rest of science, astronomy has been the beneficiary—and victim—of Moore's Law, which states that the capacity of computers and other silicon-based devices like charge-coupled devices, or C.C.D.'s, doubles every 18 months. (The C.C.D. has largely replaced photographic film in astronomical cameras.)

Projects like the National Virtual Observatory, which was created in response to the tsunami of data that is threatening to drown astronomers, is creating a new branch of science, Dr. Szalay believes.

Science, he points out, was "originally empirical, like Leonardo making wonderful drawings of nature." He continued: "Next came the theorists who tried to write down the equations that explained the observed behaviors, like Kepler or Einstein. Then, when we got to complex enough systems like the clustering of a million galaxies, there came computer simulations, the computational branch of science. Now we are getting into the data exploration part of science, which is kind of a little bit of them all."

Because its primary tools are computers rather than giant, multimillion telescopes, this new form of astronomy has the potential to democratize science.

"If at the same time most of the telescopes in the world are actually putting all of their data online with proper explanations," Dr. Szalay said, "then it doesn't matter where somebody is sitting, they can access all the data—either somebody in Baltimore, or somebody from Africa who got a Ph.D. in the U.S. and returned there and doesn't have access to a telescope but suddenly has a bunch of students. They can actually get to first-class data."

In the past 25 years the number of C.C.D. pixels in all the world's telescopes has increased by a factor of 3,000, with each pixel acting as a miniature astronomical instrument. The result, Dr. Szalay says, is that the total amount of astronomical data collected every year is doubling even while the amount spent on astronomy remains constant.

"We are getting overwhelmed," Dr. Szalay says. "With this explosion it's not just that individual telescopes are getting more and more data, but also the threshold gets lower, so that more and more groups are putting big cameras on their instruments. Even amateur astronomers today can generate gigabytes of data per night by attaching a digital camera to their telescope."

The problem is how to mine this vast store of data for the riches it almost certainly contains. Astronomers have been busy over the past couple of decades compiling complete surveys of the sky, encyclopedic catalogs of millions of astronomical objects viewed at many different wavelengths. These surveys exist in about 10 different spectral bands, from X-rays to the infrared, with each survey giving a different view of the universe.

The surveys contain about 100 terabytes (one terabyte is 1,000 gigabytes) of data, roughly five times as much as the Library of Congress holds. Unlike the Library of Congress, however, this information does not reside in a single place.

"There is no Library of Congress for astronomy," Dr. Szalay says, and as long as the data continue to accumulate at an exponential pace, there will never be one. Instead, "there will always be 8 or 10 large projects that contain 90 percent of the world's data at any one time."

The goal of the National Virtual Observatory is to make sure that "the current generation of professional and amateur astronomers are not overwhelmed by the chores of getting the actual data," Dr. Szalay says. "So we have to make it simple and easy for them to use the data in a friendly way."

In the first stage of the project this has meant creating tools that can search through different databases without requiring the searchers to be experts in their individual details. As a kind of shakedown cruise, the researchers at the National Virtual Observatory decided to focus on the data contained in two large sky surveys known as the Sloan Digital Sky Survey, which looks at the sky in the visible band of the spectrum and the Two Micron All Sky Survey, or 2MASS, which looks at the sky in the infrared.

"The reason we did those two is that they're very deep, they dig out objects that are very faint, much fainter than other surveys have been able to generate," said Dr. Bruce Berriman, a California Institute of Technology astronomer involved in the demonstration. "Because it goes to very faint objects you're able to dig out sources that are unusual or important in ways other projects can't do."

In particular, by combining the surveys they hoped to spot brown dwarfs. Brown dwarfs are essentially failed stars, lumps of matter bigger than a planet but not large enough to kindle the thermonuclear fire of a star. As a result, they are relatively cool, emit very little light and are therefore difficult to spot.

The temperature of a star, like that of a glowing piece of metal, determines the color of light that it emits: the cooler the star the redder the light. The light from the brown dwarfs that the astronomers were searching for straddles the border between the infrared and the visible. This means that a brown dwarf should appear in the very shortest wavelength band of the infrared 2MASS survey and also in the longest wavelength band of the visible Sloan survey.

An astronomer looking at just the data from, say, the Sloan survey and seeing an object in a single band would probably dismiss it.

"Chances are pretty good that that single band detection is a piece of junk, some sort of artifact in the detectors in the telescope, a glint off a bright star, any number of things," said Dr. Davy Kirkpatrick, a member of the Caltech team. But if that same object also appears in the 2MASS data then the chances shoot up that it is something worth looking at more closely.

The astronomers developed a program that could access these different databases and search them for matches. Within a few minutes the computer spit out a half a dozen or so candidates for possible brown dwarfs.

Most of these had been previously noticed in the data, which others had sifted through manually.

Finding these brown dwarfs was supposed to be the goal of the demonstration, a debugging run to prove that the software worked. But the computer also found several candidates for new brown dwarfs.

"Astronomers' first reaction when you find a new result is that there's something wrong," Dr. Berriman said. But after looking at the data more closely "it slowly dawned on us that this was something real, that this was a brown dwarf we found."

"Then our eyes started to widen up a little bit at the prospect of what might be coming in the future," he continued.

Dr. Szalay says, "This shows how many new things will come out in this process once hundreds of astronomers are using it all over the place."

What makes this result even more impressive is that the overlap of the two surveys covered something less than one two-hundredth of the sky, and yet they almost instantly found objects that astronomers, poring over data for weeks, had previously missed.

Another aspect of the National Virtual Observatory is the creation of an astronomical search engine, a kind of Google for astronomy that will allow amateurs and professionals to find astronomical resources.

"Were you to go to Google right now and type in the word galaxy, you wouldn't just get a whole bunch of astronomy sites," Dr. Berriman says. "You'd also find out about the L.A. Galaxy soccer team. There's even a town in Texas called Galaxy, with its own Web page. That's no good to astronomers because there's so much clutter in the sites."

To help solve this problem the National Virtual Observatory is creating an online registry of astronomical resources that should be available to the public early next year.

The registry, and indeed the entire virtual observatory project, is intended as a tool for anyone interested in astronomy.

"One of the major components of the registry is collecting information about the suitability of the resource for educational purposes, for amateurs, for students," said Dr. Robert Hanisch of the Space Telescope Science Institute in Baltimore. "We have a subgroup of our project concentrating on what kind of information does that clientele need to know."

The success of the National Virtual Observatory and similar projects means that exploring the heavens will no longer be limited to those few hearty individuals willing to sit freezing on mountaintops, waiting for the clouds to clear. Adding myriad seeking eyes and pondering brains to those already contemplating our place in the universe will be the greatest achievement of this new technology.

IN REVIEW

1. Explain why the amount of collected astronomical data has increased exponentially in the last few years.
2. How does the virtual observatory allow scientists and amateurs to analyze an increasing amount of collected data?
3. Explain why the virtual observatory could deepen understanding by elementary, middle, and high school students of the fundamental concepts of science.
4. How could data from the virtual observatory be used in science museums to generate interest by its visitors?
5. In your opinion, should astronomical data be analyzed by only those who have advanced degrees in science, or are there significant roles that amateur science enthusiasts can play? Do you know of any other tools besides the virtual observatory that allow amateurs to contribute to science?

Even before President Bush redefined the vision of the U.S. space program in February 2004, NASA had been moving forward with plans to send humans back to the Moon and, eventually, to Mars. Prior to sending humans, NASA will send robotic explorers similar to the Mars exploration rovers Spirit and Opportunity that will land on the Moon and Mars, and will also send orbiting platforms providing remote reconnaissance, looking for future landing sites and water. For human exploration and large robots, NASA will have to design new launch and exploration vehicles that will allow heavy payloads to go beyond Earth orbit and be transported to the Moon and Mars. In addition, to provide for a sustained human presence on the Moon and Mars, habitats protecting humans from extreme cold, unbreathable atmospheres, and harmful solar radiation will have to be designed.

New Moon: Planning the Return to Space

By Kenneth Chang
***The New York Times*, January 20, 2004**

For the future of human space exploration, NASA may turn to retro spacecraft designs.

The spacecraft that is to take astronauts back to the Moon, called the crew exploration vehicle, may turn out to be more than passingly reminiscent of the old Apollo spacecraft. NASA is not about to start rummaging about museums for leftover equipment to fulfill President Bush's mandate, but many outside experts expect the space agency to turn away from sleek, winged, reusable designs like the shuttle and return to the simple capsules that parachute back to Earth. Such designs, updated with modern electronics, would also help the agency stay within the modest budgets that Mr. Bush has set out for the Moon shot.

"The concept has been very well proven out," said Dr. Jerry Grey, director of science and technology policy for the American Institute of Aeronautics and Astronautics. "It's a rumor that was kicking around that they were going to use Apollo-type hardware."

While getting back to the Moon may be straightforward, the further goals of establishing a permanent base and then going on to Mars are more complex. The president has not offered a timetable for either, but each will require many technologies not yet invented. No one knows, for example, how to run a construction site on the dusty, airless surface of the Moon, and at most locations, a base would require power systems that persevere through the two-week-long lunar nights.

While challenging, "I think those are all feasible," said Dr. Haym Benaroya, a professor of mechanical and aerospace engineering at Rutgers University, who has studied, out of his own curiosity, the construction challenges of a lunar base.

The tasks for a Mars mission are more challenging. A robotic factory would be sent ahead to make fuel for the astronauts' return trip. Water would have to be efficiently recycled for two and a half years. Perhaps a new propulsion system would be developed to shorten the duration of the 50-million-mile trip, which would take at least

Scenes From a New Base, Reached With Existing Rockets

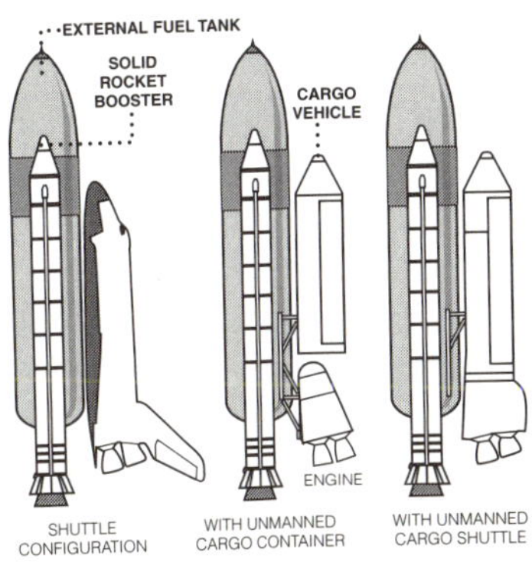

RECYCLED PROPULSION One possibility for ferrying humans and supplies to the Moon is to adapt the design of fuel tanks and rocket boosters from the shuttle program. The shuttles themselves are to be decomissioned by decade's end.

A pressurized lunar rover for exploration.

Well before President Bush's speech last week, NASA came up with this scenario for a Moon base. A lunar plant, above left, liquefies oxygen for fuel after extracting it from the soil. At right, crew members fuel their ship before the return to Earth. A conical crew module is on the ship's front.

Source: Astromaterials Research and Exploration Science Office, NASA; Greg Zsidisin, AstroExpo.com

The New York Times; images by NASA

half a year with present-day chemical rockets.

A Mars spacecraft must also be rock-solid reliable, as a rescue mission would be neither swift nor simple.

But before NASA can consider ambitious lunar or Mars missions, it must first design and build the successor to the shuttle, a task it has repeatedly attempted and failed. Since the 1980's, the agency has produced a succession of futuristic concepts that fell short of reality.

For example, Lockheed Martin in 1996 contracted to build the chubby-finned X-33, which was to blast into space like a rocket and then glide back like the shuttle. It used innovative technologies like carbon composites for its fuel tanks to cut down on weight. But the carbon composites leaked, the project went over budget and NASA canceled it in 2001.

Even the second-to-last revision of NASA's post-shuttle plans, in November 2002, retained the winged motif: the orbital space plane, which was to ferry astronauts to and from the International Space Station.

Former Senator John Glenn and other members of an advisory council for NASA asked why the orbital space plane had to be a plane rather than a capsule like the Apollo command module.

A group of experts enlisted by NASA to study the capsule idea concluded it had merit. "The command module is simple," said Dale Myers, a former NASA deputy administrator who was a member of the group. "It has an easy heat shield design. You need all the simplicity and robustness you can get."

The crew exploration vehicle now replaces the orbital space plane in NASA's plans. The exploration vehicle is to be a modular design, able to carry astronauts to the space station as well as the Moon.

NASA has said it is too early to say what it may look at. "I think we've got to avoid getting fond of a design," Sean O'Keefe, the NASA administrator, said last week at a news conference.

But Mr. Myers is certain that the part that returns to Earth with the astronauts will be a capsule, not a plane. "There's no question it's going to be a capsule," Mr. Myers said. "Wings don't do any good on the Moon. They're just dead weight."

To develop the exploration vehicle, NASA has budgeted $5 billion through 2008, when unmanned testing is to start. The spacecraft is to become operational in 2014, and total costs through 2020 will be about $30 billion, NASA estimates.

To launch the exploration vehicle, NASA no longer has anything like the Saturn 5 of the Apollo program, which could lug 280,000 pounds of payload to low Earth orbit. The space shuttle, which will be retired in 2010, can lift about 60,000 pounds, and the current generation of expendable rockets lift only about 40,000 pounds.

But not everything would have to go up in one package. The lunar version of the exploration vehicle could consist of three components—a rocket to push the astronauts out of Earth orbit, a lunar lander and the capsule to bring the astronauts back. These could be brought up separately on expendable rockets and then autonomously dock together, much as the unmanned Russian Progress supply ships dock with the International Space Station. The joined spacecraft would then depart for the Moon.

A second option would be to revive plans shelved a decade ago for a cargo version of the space shuttle. It would use the same external fuel tank and solid rocket boosters as the shuttle, but instead of an orbiter, it would have a disposable cargo container strapped on to the tank capable of lifting up to 170,000 pounds.

"That's a dimension of what we're going to have to lay out here, of how you provide cargo capabilities," Mr. O'Keefe said at last week's news conference.

Cargo capacity will become even more important when NASA starts considering what it will do once it gets back to the Moon. An agency study in 1992 looked at establishing a simple Moon base, part of the Space Exploration Initiative that the elder President Bush proposed in 1989, and even that would cost $25 billion and require the use of Saturn 5-scale rockets to transport it.

"I wouldn't call it a lunar base," said Dr. Michael B. Duke, who worked at NASA's Johnson Space Center in Houston at the time. "It was single habitat and a capability to do a few things on the lunar surface."

Under that plan, astronauts were to stay for only 45 days before returning to Earth.

A follow-up study in 1993 concluded that if oxygen could be extracted from the lunar soil for use as fuel for the return trip, costs could be reduced by $5 billion and smaller, shuttle-size cargo ships would suffice.

Since then, there has been only a smattering of research on how to build a lunar base. "One has to build in a severe environment, something we've never done before," said Dr. Benaroya of Rutgers.

Lacking both air and a magnetic field, the Moon's surface is bombarded by radiation and sand-grain-size micrometeorites hitting at tens of thousands of miles per hour. For protection, a long-term habitat would probably need to be buried under 10 feet of soil, Dr. Benaroya said.

The ubiquitous dust that covers the Moon's surface will also pose problems, clogging up machinery and wearing it out.

The first structure might be something similar to one of the components of the space station, built on Earth or in space and then taken to the lunar surface. "Essentially, it's your mobile home," Dr. Benaroya said. "That first one will be a tin can."

Later additions might essentially be balloons that arrive folded up, then inflated and stiffened, perhaps with injected foam.

The most promising location for the first lunar base appears to be an area near the south pole continually bathed in sunlight, allowing the use of solar panels for power. Large amounts of frozen water are also believed to exist in deep craters there.

In his speech, Mr. Bush suggested that the eventual trip to Mars would start at the Moon, saying, "Spacecraft assembled and provisioned on the Moon could escape its far lower gravity using far less energy, and thus, far less cost."

Dr. Robert Zubrin, president of the Mars Society, who has been pushing for a goal of sending astronauts to Mars within 10 years, said the idea of building a spacecraft on the Moon was "technically nonsense."

Dr. Zubrin, who developed a plan for a direct Earth-to-Mars mission a decade ago while an engineer at Martin Marietta, applauded the decision to phase out the shuttle and roll back activities at the space station but said he thought the priorities in Mr. Bush's proposal were reversed.

"The right way to do Mars is to design a coherent set of hardware that can do Mars," he said. Along the way, a Moon mission employing some of the Mars equipment might be a worthwhile test.

The key would be developing a rocket with the lifting power of the Saturn 5, either a cargo version of the shuttle or a larger version of the Delta 6 or Atlas 5.

Dr. Zubrin said the mission could be done in two launchings. First, an empty spaceship for the return would be sent. Once it landed it would begin to produce fuel, combining carbon dioxide from the Martian air with hydrogen to produce water and methane, and then splitting the water into hydrogen and oxygen. A year later, a spacecraft with the astronauts would arrive.

That project, he said, could be accomplished in a decade, for about $50 billion.

IN REVIEW

1. List the major challenges that need to be overcome to get humans back to the Moon.
2. The Space Shuttle, which can only achieve a low Earth orbit, is not a good prototype for a more advanced vehicle that could take humans to the Moon. Explain why the Space Shuttle is a poor prototype.
3. How could the International Space Station, which is scheduled for completion in 2010, be used in helping to get humans to the Moon and Mars?
4. Explain why finding water ice on the Moon and Mars is so important for future human exploration.
5. The cost of human exploration of the Moon and Mars will potentially be hundreds of billions of dollars. In your opinion, are these costs justified? Explain.

Since December 1972 no person has set foot on the Moon, and despite successful follow-on missions of Skylab, the Space Shuttle, and the International Space Station, NASA's crowning achievement remains the Apollo program. The Apollo missions returned hundreds of pounds of rocks, greatly increasing our understanding of the geology of both the Moon and early Earth. However, this deepened geological knowledge of the Moon failed to capture the hearts and minds of the public. Now, with NASA proposing to return to the Moon, both the agency and the scientific community are struggling with selling this renewed exploration program to the American taxpayer. Specifically, what is the use of the Moon beyond geological investigation?

The Allure of an Outpost on the Moon

By Kenneth Chang
***The New York Times,* January 13, 2004**

For some, it is the steppingstone of the Moon, not the distant goal of Mars, that is the irresistible destination in the human venture into space that President Bush will propose tomorrow.

For geologists, Moon rocks could tell much about the first billion years of Earth's history. For astronomers, the Moon would be a cold, dark place ideal for a telescope staring deep into the cosmos.

"There's a lot we could get out of the Moon," said Dr. Allan H. Treiman, a scientist at the Lunar and Planetary Institute in Houston. But he added, "It's not grab-the-public-by-the-throat science."

After the climactic triumph of Apollo, the public lost interest in continued human exploration of the Moon, and the Nixon administration cut deeply into NASA's budget. In the three decades since, NASA has focused robotic missions on more distant, more mysterious worlds like Mars and Jupiter and sent only two small orbiting spacecraft to the Moon. Clementine, a joint effort with the Department of Defense, found signs of frozen water at the lunar south pole in 1994. In 1998 and 1999, the space-on-a-budget Lunar Prospector mission, which cost $63 million, found even stronger evidence of ice and mapped out the Moon's gravitational and magnetic fields.

The primary reason to return and establish a permanent base on the Moon would be to assist a mission to Mars. Because the Moon's gravity is one-sixth of Earth's, gathering raw materials there—everything from metal for the spaceship to water for the astronauts to drink—would be much cheaper than hauling them up from Earth. So the cost and difficulty of traveling to Mars would be reduced.

The Moon base would also serve as a proving ground for new technologies developed for a Mars mission.

"If we learned to do that nearby, it might be a lot easier," Dr. Treiman said. "That's one of the big deals of using the Moon as a way station. We could learn how to live on Mars by living on the Moon."

But the Moon has its own appeal. About 4.45 billion years ago, a planetary interloper the size of Mars slammed into the infant Earth, tossing a blob of rock into space that became the Moon. With only one-eightieth Earth's mass, the Moon long ago cooled to the core, leaving it geologically dead. It is also too small to hold on to an atmosphere.

It is just this deadness that excites geologists. They see it as a museum of the history of the solar system. "To learn about the Moon is going to tell how the Earth formed," Dr. Treiman said.

On Earth, the continuous march of plate tectonics has destroyed almost all of the surface rocks from its first billion years. On the Moon, those rocks are still on the surface. The youngest rocks on the Moon are as old as some of the oldest rocks found on Earth: 3.2 billion years.

The craters on the Moon also preserve a record of the early bombardment of meteors. "The moon is a great place to study geological processes and it nicely complements the missing geological record on the Earth," said Dr. Paul D. Spudis, a planetary scientist at Johns Hopkins's Applied Physics Laboratory.

The 843 pounds of rocks brought back by Apollo astronauts revolutionized scientists' understanding of the Moon. The similar mix of oxygen atoms in the rocks of the Moon and Earth showed the two had a common ancestry instead of the Moon's forming elsewhere and then being captured by Earth's gravity. The chemical composition also showed there had never been significant amounts of water in most areas, except possibly the polar regions.

But the rocks came from just the six Apollo landing sites, leaving the rest of the surface, the size of Africa, unexplored. Dr. David S. McKay, a scientist at NASA's Johnson Space Center, would like the agency to go to other locations and dig trenches perhaps 100 yards deep to examine in detail the top layer of crushed rock and dust, known as the regolith.

"I think there is a lot of data hidden away on the Moon that remains to be unraveled," Dr. McKay said. "The lunar regolith is like a giant tape recorder that has been running for billions of years."

As astronomers try to look farther into the universe, they need a large telescope that can stay focused on a single patch of sky for weeks or months. Near absolute-zero temperatures and an airless environment are needed to prevent blurring. A nearby Moon base would allow easy repairs and upgrades.

"The Moon is a place that has been thought about a lot that has a lot of what you want," said Dr. J. Roger P. Angel, a professor of astronomy at the University of Arizona. He has proposed putting a large infrared telescope in a deep crater at the Moon's south pole.

At a Senate hearing on lunar exploration in November, Dr. Angel suggested that the mirror of such a telescope might consist of a round dish, 20 yards wide, with a reflecting liquid that spun at a rate of two revolutions a minute. The centrifugal force, coupled with the Moon's gravitational force, would push the liquid toward the outer edges of the dish to form a perfectly curved surface for gathering starlight.

Not only will a lunar telescope be more sensitive than the Hubble Space Telescope, but it should be able detect galaxies and stars far fainter than will be seen by Hubble's planned replacement. It may even pick up light from the very first stars of the universe half a billion years after the Big Bang. "That's something you could do brilliantly from the Moon," Dr. Angel said.

IN REVIEW

1. In what major ways has past lunar exploration furthered our understanding of Earth?
2. List the major ways that a Moon base would assist in our exploration of Mars. Would a Moon base be better for future space exploration than an orbiting space station?
3. What would be the major difficulties in establishing a Moon base? How would our experience in living in Antarctic scientific stations help us? How would our experience in living in the International Space Station help us?
4. Explain why the Moon would be a good location for a large telescope. Why would astronomer Dr. Angel want to put a telescope that detects infrared light in a deep crater at the Moon's south pole? Explain.
5. In your opinion, would NASA's money be better spent on extending human exploration to the Moon and Mars, or should the agency focus its resources on robotic exploration?

The Spitzer Space Telescope, launched by NASA in August 2003, is a space-based observatory that follows the very cold wake of the Earth as it orbits the Sun. Spitzer, which detects cosmic infrared light that is blocked by the Earth's atmosphere, is specially suited for penetrating cold, dusty, and gaseous regions that are hidden from optical telescopes. Regions of particular types of dust and gas, called proto-planetary disks, are an area of scientific focus for the Spitzer telescope. These proto-planetary disks, which surround nearby stars, are thought to be indicators of newly forming planetary systems. Examining such disks will help us understand the formation of our solar system much better.

Watching as Dusty Disks Slowly Turn Into Planets

By Dennis Overbye
***The New York Times,* December 21, 2004**

Astronomers think they know what goes into making a planetary system, namely dust—lots of it—swirling around a newly minted star.

So it has been encouraging that astronomers have detected and even photographed dusty disks around many nearby stars, and they have inferred the presence of more than 100 planets around other stars. But until recently they had never seen dust and planets around the same stars.

This month, astronomers said they had closed the loop.

New observations by the Hubble Space Telescope and the Spitzer Space Telescope of dusty disks around nearby stars, they said, had given them a snapshot of the entire squence by which planets coagulate out of the primordial swirl and clear out the insides of the disks, turning them into doughnuts of far-flung dust.

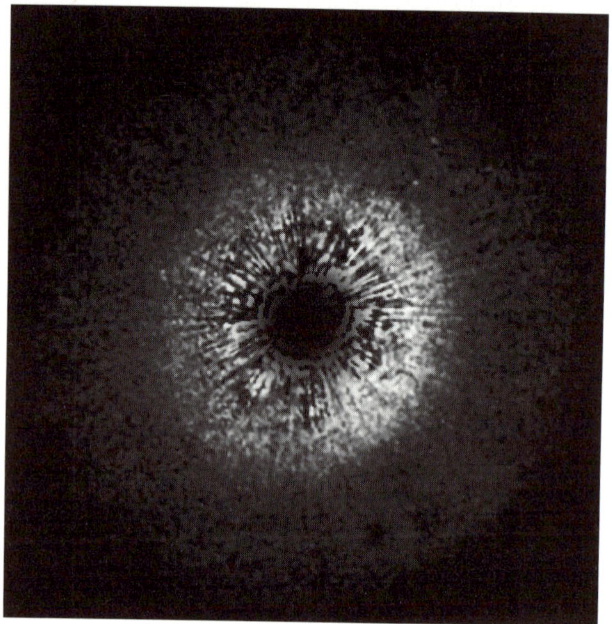

Photographs by NASA/ESA

A debris disk, more than 40 billion miles across, encircles a star.

Using Hubble, astronomers photographed thick disks of dust around a pair of young stars that might still be in the process of forming planets. Meanwhile, the Spitzer telescope trained on 26 stars known to harbor planets detected the infrared, or heat radiation, from dust surrounding 6 of the stars.

"Spitzer really closes the loop. It shows us the planets and the dust," Dr. Alycia Weinberger, a planetary expert at the Carnegie Institution of Washington, said at a news conference at the Jet Propulsion Laboratory, the base for Spitzer's control team. Dr. Weinberger compared the new observations to seeing both a house and the bricks out of which the house had been made.

One team, led by Dr. David Ardila of Johns Hopkins University, used Hubble to photograph the disk of dust surrounding the star known as HD 107146, about 88 light-years away in the constellation Coma Berenices.

The star is some 50 million to 250 million years old, young enough so that any rocky planets there may still be forming. This is the first time that a disk has been photographed around a star similar to our Sun, he said.

Viewed nearly face-on, it resembles a reddish doughnut with a hole large enough to contain the orbits of the planets in our own solar system, Dr. Ardila said.

"This picture," he said, "provides a context in which to study the history of our own solar system."

Another team, led by Dr. John Krist of the Space Telescope Science Institute, used Hubble to photograph a disk, seen edge-on, around a reddish dwarf star known as AU Microscopii. It, too, has a cleared-out space in the middle as big as the orbit of Uranus and hints of warping due perhaps to the gravitational pull of a planet.

A similar architecture has been deduced for the disks detected by the Spitzer telescope, which is designed to see long-wavelength infrared, or heat radiation, invisible to the eye.

From the relatively low temperatures of the dust, only about 50 degrees above absolute zero, the Spitzer team, led by Dr. Charles Beichman of the Jet Propulsion Laboratory was able to conclude that the disks are all far from their stars. They are on the order of 10 times the distance of Earth from the Sun, but much thinner than the disks seen by Hubble around the younger stars.

All this fits, the astronomers said, with the idea that planets form out of thick disks of gas, gradually clearing them out. The pressure of starlight also blows dust out of the system, but the constant banging and colliding of planetoids creates new dust, which winds out in a thin outer doughnut.

"Young stars have huge reservoirs of planet-building material, while older ones have only leftover piles of rubble," Dr. Beichman said in a press release from the Jet Propulsion Laboratory. "Hubble saw the reservoirs and Spitzer, the rubble."

Our own solar system, the astronomers pointed out, has a similar disk of junk, known as the Kuiper Belt, a band of icy debris on the fringes of our own solar system, as well as dust in the inner regions that gives rise to the zodiacal light. But none of this would be visible to Spitzer or Hubble, the astronomers say.

By correlating the details of the architecture of these systems with the planets that live there, and with future observations by more sensitive spacecraft, the astronomers said they hoped to find out more about what makes planetary systems tick.

"These new disks discovered by Spitzer have the potential to be our Rosetta stone," Dr. Weinberger said.

IN REVIEW

1. Why are data from both the Hubble and Spitzer space telescopes needed to tell a more complete story of planetary formation?
2. Astronomers are interested in looking at the dusty regions before and after planets have been formed. Explain why this would be of interest to them.
3. Why would newly forming planets create a "doughnut hole" around the stars that have been observed by Hubble and Spitzer? What other explanations could account for this hole besides newly forming planets?
4. Explain why Hubble is better suited for detecting dust clouds that are the "reservoirs" or building blocks for planets, and Spitzer is better suited for detecting the rubble that is left after planets are formed.
5. Discuss how all of these observations are needed to support scientific theories of planetary system formation. How might these new Hubble and Spitzer observations lead to alteration of these theories?

Stars are massive objects with nuclear fusion occurring in their cores. Failed stars, more commonly called brown dwarfs, are objects that are 10 to 100 times more massive than Jupiter, but not massive enough to produce long-term fusion. Because brown dwarfs are relatively cool (around 1000 K on their surface), they are very dim and difficult to spot with optical telescopes. However, brown dwarfs may be very plentiful and may account for much of our galaxy's mass. In addition, new observations from the Spitzer Space Telescope show that planetary systems may form around brown dwarfs, raising the possibility that some of these planets may be warm enough for liquid water to exist on their surface and potentially be life-supporting.

Dusty Planet-Forming Process May Be Playing Out in Miniature

By Dennis Overbye
***The New York Times,* February 8, 2005**

Astronomers have found what they say may be a miniature solar system in the making.

In the latest example of the apparent fecundity of nature, a tiny starlike object too small to be a star seems to be surrounded by a tiny disk of dust that could someday form planets, which could perhaps even be briefly capable of supporting life, according to observations.

The discovery raises the possibility that astronomers will soon discover actual planets and habitable abodes around objects that are barely larger than planets themselves. The starlike body, only about 15 times as massive as the planet Jupiter, is the smallest object yet to be found with such a disk. With a temperature of 3,600 degrees Fahrenheit, it is also the coolest. As such, the object is not really a star at all, but a failed star known as a brown dwarf that is only barely bigger than the largest giant planets.

"It's just incredible," said Dr. Kevin Luhman of the Harvard-Smithsonian Center for Astrophysics. "A little disk that could form planets around an object that is small enough itself to be a planet." Dr. Luhman led an international team of astronomers that observed the brown dwarf, known as OTS 44, using the Spitzer Space Telescope. He will be discussing his results this week at a meeting on extraterrestrial planets at the Aspen Center for Physics in Colorado. A paper describing the results will be published in The Astrophysical Journal Letters.

Photo by NASA/J.P.L.

An artist's conception of the disk of gas and dust that was detected by the Spitzer Space Telescope around the brown dwarf OTS 44.

Dr. Geoffrey W. Marcy of the University of California, Berkeley, said the report opened a new possible abode for life, on Earth-like planets orbiting brown dwarfs. "While these earths would be warmed up for only a few million years," he said in an e-mail message, "it is interesting to imagine what biochemistry might flicker valiantly during that brief period when the brown dwarf is luminous enough to warm up the earth, only to freeze over before Darwinian evolution can kick in."

The work continues a recent trend in which astronomers using bigger and bigger telescopes have shifted their attention somewhat, from the biggest and brightest objects in the cosmos—supernova explosions, quasars and giant galaxies—to the smallest and dimmest. Little dim things far outnumber their more striking and violent cousins, and it is among these humbler denizens of the cosmos that astronomers must look for life or life-friendly conditions.

The Spitzer Space Telescope, launched into orbit around the Sun in 2003, is the fourth and last of NASA's so-called Great Observatories. It is designed to see the infrared, or heat radiation emitted by celestial objects.

Cool dim objects like small stars, brown dwarfs and planets emit electromagnetic radiation most strongly in infrared wavelengths, which are longer than those of visible light, and which can pass through interstellar dust clouds that often shroud new stars, but Earth's atmosphere blocks infrared light. And so prospective infrared astronomers have had to go to space, starting with the Infrared Astronomical Satellite, IRAS, in 1983, which discovered disks of dust around several stars, including Vega.

Since then disks have been detected or have been inferred to exist around thousands of stars, including a few brown dwarfs. In December astronomers using Spitzer announced that they had detected the signatures of dusty disks around half a dozen stars already known to harbor giant planets, thereby closing the loop between disks and planets.

By searching out the smallest dwarts that have such disks, Dr. Luhman and his team hoped to learn more about how brown dwarfs formed and determine whether planets could exist around such objects.

The brown dwarf, OTS 44, is part of a cluster of very young stars about 500 light-years away in the constellation Chamaeleon. When the astronomers looked at it with the Spitzer, they found that the brown dwarf was putting out excess heat at certain wavelengths, a signature of cooler material surrounding the brown dwarf.

The extra heat, they calculated, was coming from a disk of dust and perhaps larger materials, extending 300,000 to 900 million miles from the dwarf star, about a tenth the size of the disk that formed our own solar system. They estimated that there was enough material in the little disk to make a planet like Jupiter or several smaller Earth-like planets.

The disk around the brown dwarf is unlikely to have any planets, Dr. Luhman said; based on the ages of the stars in the cluster, the brown dwarf and its disk are only about a million years old. It could take 10 million years for the small bits and chunks of rock and dust to coagulate into planets.

But he added, "There could be little embryos we can't see."

A planet that orbited close enough to the brown dwarf could even be briefly habitable, meaning that liquid water could exist on it, Dr. Luhman said. This "habitable zone," as astronomers call it, would extend from one million to four million miles from the brown dwarf, well within the confines of the disk, according to theory.

But as the brown dwarf cools, Dr. Luhman explained, this zone will sweep in, quickly at first and then ever more slowly. That means that the fate of a planet depends on where it is. A planet in the center of the zone now could look forward to a few tens of millions of years of habitability, he said.

But he added that theorists had no idea how and if such planets would form to begin with. Nor do they know whether life could arise in the brief time allotted.

Other astronomers said it was a little too early to start speculating about life or prebiotic chemistry around a brown dwarf.

Dr. Alan Boss, a planetary theorist at the Carnegie Institution of Washington, said, "The biggest impact of this discovery may be to force theorists to consider the possibility of planet formation around objects which are so low in mass that they are on the verge of being called planets themselves."

It is not too soon, he said, to start worrying about what to call some of these things.

Four years ago, astronomers discovered a dozen or so objects with masses estimated at 5 to 15 Jupiters amid new stars and gas clouds in the Orion nebula. They named them free-floating "planets," to the dismay of most astronomers, who object that planets are things that revolve around stars. A committee of the International Astronomical Union came up instead with the name "sub-brown dwarfs."

The confusion has been compounded lately by the recent "weighing" of a putative brown dwarf in the constellation Doradus, which suggest that some of these cosmic small fry, like brown dwarfs, might be more massive than they look. If this is true, the free-floating planets may actually be brown dwarfs, and some brown dwarfs may actually be small stars.

Dr. Laird Close of the University of Arizona, who made those measurements, said that even given the uncertainty in the brown dwarf's mass, OTS 44 was the smallest object yet found with an infrared excess—and thus probably a circumstellar disk. But in an e-mail message, he cautioned that this did not prove that 15-Jupiter-mass objects could form their own solar systems.

Dr. Boss, chairman of the astronomical union's committee, said, "More classification headaches are likely to ensue as astronomers probe deeper and deeper into the backyards of the Sun's neighborhood."

IN REVIEW

1. Explain why the Spitzer Space Telescope, which detects infrared light emitted from objects, is able to detect brown dwarfs much more easily than the Hubble Space Telescope.
2. What was the specific evidence for a dust disk around the brown dwarf? How strong is the evidence for planets around brown dwarfs?
3. What sets this observation apart from other observations of dust disks?
4. How is the "habitable zone" around this brown dwarf different from the habitable zone around a normal star? Consider both the location and the longevity.
5. Do you agree with the "other astronomers" who feel that it is too early for Dr. Marcy to bring up the possibility of life on planets orbiting brown dwarf stars? Explain.

To date, more than 150 planets have been discovered outside our solar system. These planets, which orbit other stars, are often called extrasolar planets. Until recently, all of the planets have been observed indirectly, through methods discussed in the textbook. This article discusses direct detection of extrasolar planets.

Alien Planets Show Themselves for First Time

By Dennis Overbye
***The New York Times,* March 23, 2005**

Astronomers said yesterday that they, or at least their telescopes, had laid eyes for the first time on planets beyond the solar system.

Using the NASA Spitzer Space Telescope and careful timing, teams studying two planets were able to distinguish the glow of the planets' infrared radiation from the overwhelming glare of their parent stars. Both planets are so-called hot Jupiters, massive bodies circling their stars in tight, blowtorching orbits and probably unfit for the kind of life found on Earth.

Until now, astronomers could infer the existence and some properties of these and other so-called exoplanets only by indirect means. They said directly measuring light from the planets was a major step in the quest to understand what alien planets are made of, because different molecules in the atmosphere absorb infrared light in characteristic ways and allow scientists to compare these alien planets to those in the solar system. Ultimately, astronomers would like to know if Earth, with its ability to evolve and support life, is unique or common in the universe.

Dr. David Charbonneau of the Harvard-Smithsonian Center for Astrophysics, who led one of the teams, said he was ecstatic when he first saw the data. "We've been hunting for this light for almost 10 years, ever since extrasolar planets were first discovered," Dr. Charbonneau said.

The Harvard-Smithsonian team and the other team, led by Dr. L. Drake Deming of the Goddard Space Flight Center, announced their results at a news conference at the headquarters of the National Aeronautics and Space Administration in Washington. The Deming team's paper was published online yesterday by the journal Nature; the other team will publish its results in The Astrophysical Journal on June 20.

Dr. Geoffrey W. Marcy, a planet hunter at the University of California, Berkeley, called the results "the stuff of history books" and added, "With this result, we are closer to understanding our own human roots, chemically, among the stars."

Dr. Alan P. Boss, a planetary theorist at the Carnegie Institution of Washington, said in an e-mail message that the discoveries showed "that we are well along the way to combining astronomy and biology into the new science of astrobiology, with the ultimate goal being to search for life beyond Earth."

About 130 planets are known to orbit other stars, but until now they have all been detected indirectly, either by the wobble their gravity induced in the motions of their parent stars or by slight dips in the stars' light when their planets passed in front of them.

The planets whose discoveries were reported yesterday fall in the latter category, passing directly in front of their parent stars periodically and then behind them. The Goddard team's object, known as HD 209458b, was the first planet detected by the so-called transit method, back in 1999. It circles a star about 153 light-years away in the constellation Pegasus, making a complete trip every 3.5 days. In 2001, Dr. Charbonneau used the Hubble Space Telescope to make spectroscopic measurements of this star while the planet was passing in front of it and found that sodium and hydrogen were present in the planet's atmosphere.

The other planet, known as TrES-1 after the telescope network that discovered it, the Trans-Atlantic Exoplanet Survey, is about 500 light-years away in the constellation Lyra. Discovered last year by a telescope only four inches in diameter, it circles its parent every three days.

In both cases the astronomers were able to use the special geometry of these planetary systems to tease out the faint light of each planet from its parent's glare, comparing measurements made while the planet and star were both visible and while the planet was hidden behind the star. It helped that the Spitzer telescope, launched in 2003, was built to study infrared radiation, which has a longer wavelength than visible light. Because they are so close to their stars, these planets were expected to be hot and therefore show up more easily at infrared wavelengths than in the visible light. While the stars outshine such giant planets by a factor of 10,000 or so in visible light, they are only about 400 times brighter in infrared.

When the astronomers put on their "infrared goggles," as Dr. Charbonneau put it, the planets popped into view.

The Spitzer telescope registered a dip in light of about a quarter of 1 percent in each case when the planet went behind its star.

That dip was enough to allow the astronomers to take the temperature of these planets and confirm that they are hot, as predicted: about 1,450 degrees Fahrenheit for TrES-1 and about 1,570 degrees for HD 209458b.

Other results were more puzzling. The TrES-1 planet seems to be emitting less light at shorter infrared wavelengths than the models of such planets predict, Dr. Charbonneau said, perhaps because greater-than-expected amounts of carbon monoxide are absorbing those wavelengths.

But as Dr. Charbonneau said, theoretical predictions about extrasolar planets have turned out to be "nothing short of disastrous." The field, he said, is driven by data, and there has not been much of that until now.

Dr. Boss said the Spitzer telescope had been designed decades ago and was never intended to study extrasolar planets. He also said that NASA had a whole menu of "wonderful" planet-finding missions.

"Just think of the results we will have from NASA's future telescopes that will be specifically designed to detect and study extrasolar planets," he said.

IN REVIEW

1. Describe how scientists first observed an extrasolar planet directly.
2. Explain why scientists desire to observe extrasolar planets directly. What is the ultimate question many scientists wish to answer?
3. Astronomers still used some indirect measurements when making the direct observation. Which indirect method did they use? How did this assist scientists in making a direct observation?
4. What questions arose because of the direct observation? Explain.
5. What limitations does the Spitzer Space Telescope have in its ability to detect extrasolar planets? In your opinion, should NASA develop a telescope that is able to directly study extrasolar planets? Explain why or why not.

Sending missions to Mars is very risky. To date, the United States, Russia, Japan, England, and the European Community have sent 34 missions to the red planet, with 22 failing due to launch accidents, equipment malfunctions, or crash landings. Partly because of this dismal record, the public experiences a huge amount excitement and anticipation when a mission is getting ready to land on Mars, and when the landing is successful this excitement turns into jubilation. In January 2004, this occurred with the successful landing and operation of the two Mars exploration rovers, Spirit and Opportunity. Indeed, these twin robotic explorers exceeded their designers' expectations by operating far beyond their expected lifetimes, extending their contributions to our understanding of the history of liquid water on Mars.

Unexpectedly, the Mars Rovers Are Still Going Strong

by Kenneth Chang
***The New York Times,* March 22, 2005**

Nearly a year past its planned three-month lifetime, the Mars rover Spirit has found itself rejuvenated and is now making some of its most significant discoveries about Mars' waterlogged past.

Spirit and its twin, Opportunity, on the other side of Mars have continued working so well that managers have requested that the mission be extended up to another 18 months. NASA reported that one of the mineral-identifying instruments on Opportunity had been turned off because of a malfunction, but the rovers appear otherwise healthy.

"Today is Sol 427 of our 90-Sol mission to Mars," Dr. Steven W. Squyres, the mission's principal investigator, said Wednesday, to laughter and applause of scientists at the Lunar and Planetary Sciences Conference here on the outskirts of Houston. "Things continue to go well."

A Sol is a Martian day, slightly longer than an Earth day at 24 hours, 39 minutes and 35 seconds.

Even serendipity is aiding Spirit as it roams around Gusev Crater, a 95-mile wide depression that looks as if it may have once been a lake.

To even out the wear and tear on the rovers, mission managers having been driving them backward much of the time, and as Spirit was driving backward up the hill, "We hit a break in slope, and we kind of churned," Dr. Squyres said. "I was very disappointed when I saw these images, because it's all churned up, and we didn't complete our drive. And then we looked carefully and noticed that soil looked awfully bright."

That turned out to be various types of salt, presumably left as water evaporated on ancient Mars. Opportunity was the first to find such salts last year, but "the record for the saltiest place on Mars now belongs to Gusev Crater," Dr. Squyres said. "This stuff is more than 50 percent salt."

When the rovers arrived in January last year, luck favored Opportunity, which ended up in a tiny crater and almost immediately found evidence of past water of Mars.

Meanwhile, a computer programming glitch waylaid Spirit for a week and a half, and when it resumed its work it found the rocks around it to be made of unremarkable lava. It then made a two-month, two-mile trek to hills that appeared older and more geologically intriguing.

Once it arrived at the hills, Spirit also started finding evidence of past water.

"In many ways, the Spirit mission began on Sol 156," Dr. Squyres said. "We're seeing a bewildering diversity of composition in the Columbia Hills." They were named in memory of the Columbia space shuttle astronauts who died in 2003.

Some of Spirit's aches of age have also gone away. In June, the motor driving its right front wheel started drawing two to three times as much electrical current and appeared in danger of burning out. Mission controllers limited the use of the wheel, driving on the other five. Then, three months ago, the motor returned to its normal functioning.

Dust gathering on Spirit's solar panels had cut the amount of energy they generated each day from 900 watt hours initially to just over 400. "The death zone for this vehicle is about 280," Dr. Squyres said.

As Spirit made its way up through the hills, mission controllers had to be careful that the panels stayed tilted toward the Sun. Then on March 9, the power output suddenly jumped up to 800 watt-hours, nearly as good as new. Spirit's cameras detected several dust devils—mini-tornadoes—in the area, and apparently one passed directly over the rover, blowing away the dust.

"We are back to power levels we were at on Sol 30," Dr. Squyres said.

Opportunity experienced a similar cleansing of its solar arrays last year.

After exploring the small crater it landed in and then a larger stadium-size one, Opportunity is now making a long drive across the Martian surface. It stopped to examine pieces of its discarded heat shield.

"Then it's become drive like crazy," said Dr. Raymond Arvidson, a professor of earth and planetary sciences at Washington University in St. Louis and a member of the science team. On Thursday, Opportunity drove 624 feet in one day, the longest drive to date.

IN REVIEW

1. List the major scientific discoveries made by the rovers Spirit and Opportunity.
2. Explain how the extended lives of the rovers have enabled scientists to deepen their understanding of Mars.
3. How have the scientists and engineers operating Spirit and Opportunity responded to technical glitches, enabling the rovers to continue the data collection? How has luck come into play?
4. As robotic explorers, what advantages do the rovers have over human explorers? What advantages would human explorers have over the rovers?
5. The total cost of both Spirit and Opportunity was just under $1 billion. In your opinion, is the cost justified given the scientific discoveries that they have made? Does the extended lifetime of the rovers make their costs more or less justifiable? Explain.

One-sixth the size of Earth, Mars is a frigid planet with an extremely thin atmosphere and no liquid water on its surface. Yet, of all the planets in the solar system, Mars is the planet most like our home. Its dry land area is roughly equal to that of Earth, and like Earth, Mars has volcanoes, canyons, dry riverbeds, and polar caps of water ice. Indeed, one of the most important tools that scientists use to understand Mars is direct comparisons with Earth. The Mars exploration rovers, which have returned several daily images from the surface of the planet, have provided scientists with a huge amount of data. By comparing these images to what we know from the Earth's surface, scientists have been able to deepen their understanding of Martian geological history.

Mars' Round, Smooth Stones Have a Counterpart in Utah

By Kenneth Chang
***The New York Times,* June 22, 2004**

Among the most intriguing discoveries from Mars this year are the almost perfectly round pebbles, nicknamed blueberries, that the rover Opportunity found strewn about at its landing site.

Similar spherical pebbles, it turns out, are also strewn around southern Utah.

To better understand the geological history of Meridiani Planum, the vast plain near the Martian equator that Opportunity is currently exploring, scientists are looking for places on Earth that might have formed under comparable conditions.

Researchers from the University of Utah, the National Institute of Aerospace Technology in Madrid, and G. d'Annunzio University in Italy found the Utah pebbles in several places, including Zion and Capitol Reef National Parks. They say the Utah pebbles, which formed about 25 million years ago when iron-rich groundwater percolated through porous sandstone, provide more evidence that the Meridiani Planum region was once drenched in water and perhaps amenable to life.

"It's hard to generate that kind of shape with other mechanisms," said Dr. Marjorie A. Chan, chairwoman of the geology and geophysics department at the University of Utah and the lead author of an article in the current issue

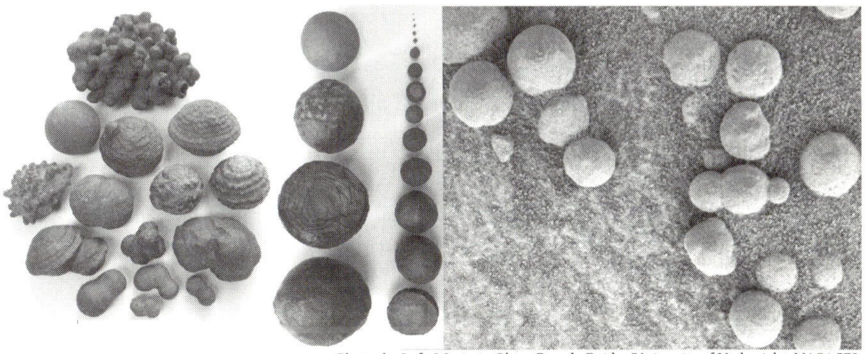

Photos by Left, Marjorie Chan, Brenda Beitler/University of Utah; right, NASA/JPL

Spherical pebbles strewn around southern Utah, left, which formed about 25 million years ago, are similar to the round pebbles found this year on Mars, right.

of Nature that compares the Utah pebbles with their Martian counterparts.

Planetary scientists chose Meridiani Planum as the destination for the Opportunity rover because it was there that orbiting spacecraft had spotted a vast deposit of hematite, an iron mineral that usually forms in the presence of water. From orbit, however, scientists could not tell which rocks contained the hematite or how the hematite got there.

Dr. Chan said that before Opportunity landed, she had predicted that the hematite would be found in spheres like those she had studied in Utah for eight years. When Opportunity did indeed find blueberries, "we were just amazed, oh my gosh it's true," she said.

The Utah blueberries were scattered on the surface and also embedded within bedrock. On Mars, the rover's probes showed that the blueberries, but not the surrounding rock or soil, contained bountiful amounts of hematite.

Scientists working on the rover mission concluded that the blueberries were objects known as concretions. Concretions form out of minerals that precipitate out of iron-rich water into surrounding sedimentary rock; because the composition of the rocks was uniform, the concretions grew equally in all directions into spheres.

Dr. Scott M. McLennan, a professor of geosciences at the State University of New York at Stony Brook and a member of the mission science team, said the Utah findings were interesting and potentially useful. "When you look at the blueberries concentrated on the surface on Utah," he said, "they look remarkably similar to the concentrations of blueberries on the Martian surface."

The Utah pebbles are not a perfect match, however. They come in a wider range of sizes, as large as grapefruits or even basketballs; the ones on Mars range from the size of BB's to small blueberries, all less than a quarter-inch in diameter. That may be because Opportunity has examined only a small area, or it may reflect a difference in how the concretions formed.

In other respects, Meridiani Planum more resembles other places on Earth. On Mars, the surrounding sedimentary rock is made of ground-up volcanic basalt like that found on the beaches of Hawaii; the Utah concretions formed within sandstone. The water at Meridiani Planum was also highly acidic, like the Rio Tinto in southwestern Spain, whose name translates as Red River from the rusty hue it acquired from iron washing out of the rocks.

Dr. McLennan that geologists would have to look at many places on Earth to piece together the story of what happened at Meridiani Planum billions of years ago. "My guess," he said, "is that's going to be hard to find a perfect analog."

IN REVIEW

1. List how scientists have used comparisons with Earth to better understand the geological history of Meridiani Planum.
2. How do the spherical pebbles imaged on Mars provide evidence of past persistent liquid water on the surface of the planet?
3. Explain how spacecraft that orbit and/or land on Mars complement each other to aid scientists in understanding the planet, particularly with regard to the issue of minerals that form in water.
4. What are the significant differences between the spherical pebbles imaged on Meridiani Planum and those found in Utah? In your opinion, do these differences invalidate the scientists' conclusions about past persistent surface water on Mars?
5. How might human exploration of Mars provide more efficient and concrete data than those provided by the robotic rovers?

Until the mid-1960s, scientists believed that Mars could be a planet with liquid water on its surface. To the disappointment of many, the Mariner, Viking, Pathfinder, Global Surveyor, and Odyssey missions to Mars showed that no flowing water is on the surface. However, these missions did show that plentiful amounts of water ice exist at the poles and that small amounts of water vapor are in the Martian atmosphere. These missions also showed evidence of ancient flowing water with images strongly suggesting that liquid created channels through erosion. In 2004, NASA sent the Mars exploration rovers Spirit and Opportunity to confirm this evidence and to provide some clues as to the date when liquid water may have existed on the Martian surface.

Scientists Report Evidence of Saltwater Pools on Mars

By Warren E. Leary
***The New York Times,* March 24, 2004**

Mars was once a much warmer, wetter place, with pools of saltwater that sometimes flowed across the surface, scientists reported Tuesday.

Analyzing findings from sedimentary rocks explored by the rover Opportunity, the scientists said the rocks now appeared to have formed under a shallow bed of softly flowing water near a shoreline—not, as formerly seemed possible, through seepage from underground.

It was the first concrete evidence that water might have flowed on the Martian surface, and it provided new hints that life may have existed there.

"We think Opportunity is now parked on what was once the shoreline of a salty sea on Mars," Dr. Steven W. Squyres of Cornell University, principal investigator for the science payload on the Opportunity and its twin Mars exploration rover, Spirit, said at a news conference here at NASA headquarters.

"If we are correct in our interpretation, this was a habitable environment," Dr. Squyres went on. "It's a salt flat. These are the kinds of environments that are very suitable for life."

He said there was still no evidence that life existed at the site, and the Opportunity does not have the instruments to hunt for microscopic fossils. But rock formed from sediments from evaporating standing water, like those at the site, "offers excellent capability for preserving evidence of any biochemical or biological material that may have been in the water," he said.

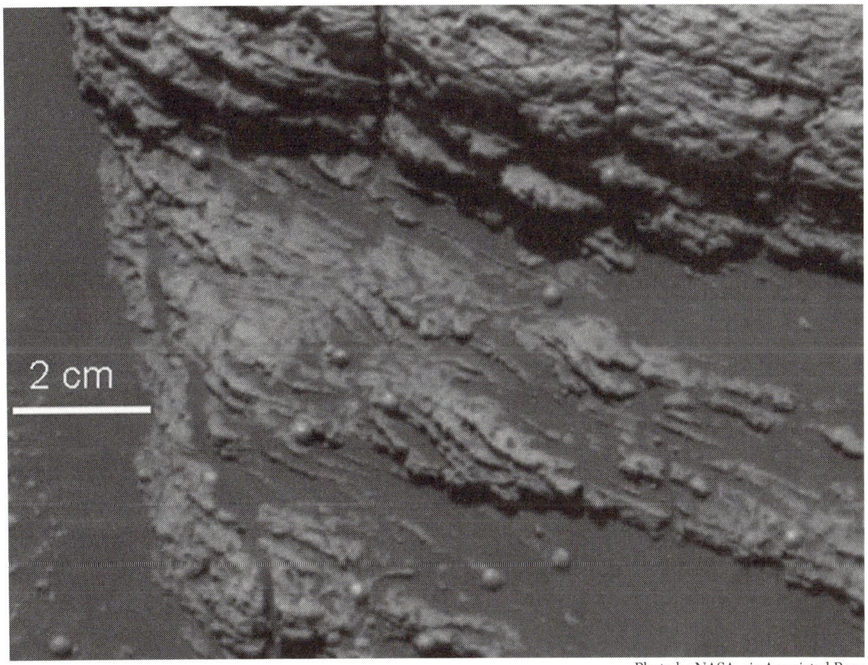

Photo by NASA, via Associated Press

This image of a rock taken by a Mars rover is said to provide evidence that sediments forming the rock were laid down by flowing water.

The rover Opportunity landed on a plain called Meridiani Planum on Jan. 25, three weeks after the Spirit touched down on the opposite side of the planet at a site called Gusev Crater. NASA officials said Tuesday that the two craft were operating so well that their original 90-day missions would very likely extend into the summer.

Dr. Edward Weiler, NASA's associate administrator for space science, said that when researchers first announced on March 2 that the Opportunity had found rocks that formed in the presence of water, there was some indication that the water pooled on the surface. But to be sure that they were interpreting the data correctly, Dr. Weiler said, they sent it to several outside experts for review before Tuesday's announcement.

One of those scientists, Dr. David Rubin, a sedimentary expert with the United States Geological Survey in Santa Cruz, Calif., said he was shocked when he received the Mars pictures from NASA. "I was astonished," Dr. Rubin said. "They looked like sedimentary deposits found at a beach on Earth."

Dr. Rubin told the news conference that some of the pictures showed rocks that could have been formed by particles deposited by the wind, and others by sediments put down by water. However, he said, there are features in some of the formations that require flowing water, which supports the explanations by the rover scientists.

Dr. John Grotzinger of the Massachusetts Institute of Technology, a member of the rover science team, said the rocky outcroppings examined by the Opportunity have ripple patterns and salt concentrations that are telltale signs of rock formed in standing water that sometimes flowed. This would be like some salt flats on Earth, which are periodically submerged, Dr. Grotzinger said.

To examine the patterns in the Mars outcroppings, the Opportunity used a microscopic camera on its robot arm to take multiple postage-stamp-sized pictures that were put together into mosaics to examine the layering in the rock.

The close-up pictures revealed that the sediments that bonded to form the rocks were in uneven layers distorted by the ripples of flowing water, patterns called crossbedding and festooning, Dr. Grotzinger said.

Patterns in some of the layered rocks indicate they were shaped by ripples at least two inches deep, and possibly much deeper, flowing at a speed of 4 to 20 inches per second, he said.

The patterns include distinctive, upward-turning curves characteristic of water depositing particles in layers rather than wind, he said.

In further evidence of water formation, Dr. Grotzinger said, the rover found chlorine and bromine salts in the rocks, suggesting that salt concentrations were rising while water was evaporating.

Dr. Weiler said the findings would spur efforts to expand Mars exploration and look for evidence of past or current life. He noted that NASA planned to send the Mars Science Laboratory, a more sophisticated rover that will do detailed life-science testing, to the planet in 2009 and that Meridiani was now a prime site.

To search for further evidence of Martian water, the Opportunity has climbed out of its landing site and will head for a crater named Endurance, 2,300 feet away. That crater, scientists said, may have even deeper rock formations.

IN REVIEW

1. What are the major lines of evidence that liquid saltwater once existed at Opportunity's landing site?
2. Explain how NASA assured that their observations and theories about past liquid water on Mars were robust and sound.
3. What scientific measurements did the Opportunity rover make to enable scientists to make their conclusions about the landing site?
4. Might the rover's findings affect future missions to Mars? In your opinion, should NASA continue to explore Mars given the fact that liquid oceans may have dried up over 3 billion years ago? Explain.

The atmosphere of Mars is very thin, just 1% of the density of Earth's. Made up mostly of carbon dioxide, the Martian air contains virtually no oxygen. With this lack of oxygen and very low pressures, Mars' atmosphere is toxic to humans, and future astronauts will require special suits as they roam the surface. Despite these differences, there are some key similarities between the atmospheres of Earth and Mars. For example, Earth and Mars have clouds and fog consisting of water ice and vapor. Recently, scientists have discovered another likeness of Mars to Earth: trace amounts of methane in the air of Mars, a potential indicator of bacterial life below the surface.

Methane in Martian Air Suggests Life Beneath the Surface

By Kenneth Chang
***The New York Times,* November 23, 2004**

A third team of scientists has now reported a seemingly simple discovery on Mars: its atmosphere contains methane.

But that finding has potentially profound implications, including the possibility of present-day microbes living on Mars.

Speaking this month at the American Astronomical Society's Division for Planetary Sciences meeting in Louisville, Ky., Dr. Michael Mumma, a senior scientist at NASA's Goddard Space Flight Center in Greenbelt, Md., reported three years of observations had provided strong evidence for methane.

"We are 99 percent confident," Dr. Mumma said. "It surprised all of us, actually. We really are still scrambling to understand what it means."

Methane, the simplest of hydrocarbon molecules with one carbon and four hydrogen atoms, is fragile in air and easily broken apart when hit by ultraviolet light. Calculations indicate that any methane in the Martian air must have been put there within the past 300 years.

That then raises the question: What is putting methane into the Martian air?

There seem to be only two plausible explanations. One is geothermal chemical reactions involving water and heat like those that occur on Earth in the hot springs of Yellowstone or at hydrothermal vents on the bottoms of oceans.

Photo by European Space Agency via Associated Press

Meandering river channels like this in Mars' southern hemisphere helped give rise to the notion that the planet had once been a tropical paradise, warmed by a heat-trapping blanket of carbon dioxide in its atmosphere.

That would intrigue planetary geologists. Although frozen water is known to exist, there are no signs that any volcanism has occurred there for millions of years. Also, an instrument aboard NASA's Mars Odyssey looked for warm spots on Mars' surface and did not find any.

The other, more intriguing, is life. On Earth, a class of bacteria known as methanogens breathes out methane as a waste product. The discovery, if confirmed, suggests that perhaps Martian life arose on a presumably more hospitable Mars billions of years ago and survives to this day underground, beneath the cold, dry landscape.

Dr. Vladimir Krasnopolsky of Catholic University in Washington, the leader of one of the teams, said he believed bacteria to be the "most plausible source."

Others are more cautious. "Three difficult detections, or marginal detections, don't equate to one really strong one," said Dr. Philip R. Christensen, a professor of geological sciences at Arizona State University.

Dr. Krasnopolsky's findings, relying on observations from the Canada-France-Hawaii Telescope in Hawaii, were first reported at a conference in Europe this year and will be published in the journal Icarus.

In January, scientists working on the European Space Agency's Mars Express mission also reported the detection of the methane. A few months later, that group, led by Dr. Vittorio Formisano of the Institute of Physics and Interplanetary Science in Rome reported that the methane appeared to be more plentiful in regions where frozen water is known to exist underground.

All three teams of astronomers looked for methane molecules in the Martian air by examining the rainbow of light reflected by the planet. Different molecules absorb different, very specific colors, producing a bar-code-like series of black lines blotting out part of the rainbow spectrum. The widths of the lines tell the quantity. Dr. Krasnopolsky and Dr. Formisano based their claims on a single dark line.

The journal *Science* published the Mars Express results this month. Dr. Christensen of Arizona State said he was unconvinced by it. "I must confess I'm surprised it was published," he said. "I think it's just instrument noise. This detection is right at the noise level of the instrument."

Dr. Mumma said his ground-based observations from Hawaii and Chile spotted two separate dark lines corresponding to methane and performed other checks. "Mike's a really careful guy," said Dr. Steven W. Squyres, principal investigator for the rovers now on Mars, who attended Dr. Mumma's talk. "It was to me, by a significant margin, the most compelling argument that I've seen."

There is a new wrinkle in Dr. Mumma's findings: some regions of Mars near the equator possess surprisingly high levels of methane, up to 250 parts per billion, while areas near the poles had 20 to 60 parts per billion. Earth air, by comparison, contains about 1,700 parts per billion of methane. Dr. Mumma's readings are considerably higher than those reported by the other two groups.

Scientists have generally thought that methane, if present, would quickly distribute evenly through the atmosphere, so the clumps of high concentration suggest that not only are there sources emitting methane, but perhaps some process is also destroying methane over the poles.

The methane findings on current-day Mars come as planetary scientists are again rethinking their ideas about long-ago Mars. Geological carvings on the surface, from ones that look like meandering river channels to gigantic canyons, gave rise to the notion that Mars had been a tropical paradise, perhaps warmed by a thick heat-trapping blanket of carbon dioxide in its atmosphere.

But climatologists found that it was hard for their computer models to provide that much warming, and scientists shifted to a picture of Mars as wet, but cold. Many of the features could have been cut by glaciers or transitory hellish deluges when ice was melted by meteor strikes.

Mars also possesses few carbonates, the minerals in limestone that would be expected to form in the presence of water, but does have much olivine, a mineral that falls apart when exposed to moisture.

This year, however, the rover Opportunity, which landed at a site called Meridiani Planum, found minerals and salts that indicate that that part of Mars at least had once been soaked in water, although when and for how long remain uncertain. Dr. Squyres also noted that while the minerals indicate liquid water, "We see nothing that looks like wave ripples" in the layers of sediments preserved in the rocks.

The other rover, Spirit, on the other side of Mars, initially found only volcanic rocks that appear almost unchanged for billions of years. It has since rolled to nearby hills, which appear to be slightly older, where the rocks seem to have been significantly changed by water.

The rover findings and others presented last month in Jackson Hole, Wyo., at a conference about early Mars have led some to think again of the planet long ago as warm and wet.

Even Dr. James F. Kasting, a climatologist at Penn State whose models helped convince people that Mars had not been warm, has changed his mind. Dr. Kasting is now investigating methane, a more potent greenhouse gas than carbon dioxide, as a cause of warming. His initial simulations show methane cooling the planet but he thinks the error is in his calculations, not his hypothesis.

"I think it's our problem, not Mars' problem," he said. "I think the evidence keeps mounting that it was warm. I think it has to be stably warm."

The opinion is not unanimous, but the idea of early oceans is gaining favor. Some scientists, like Dr. Stephen M. Clifford of the Lunar and Planetary Institute in Houston, said that four billion years ago the decay of radioactive elements in the core of Mars would have produced enough heat to melt ice from below, producing an ice-covered ocean. Acidic waters could explain the lack of carbonates.

Dr. Daniel J. McCleese, chief scientist for Mars exploration at the NASA's Jet Propulsion Laboratory, said that during discussions someone said, "So we all believe there were oceans on early Mars?"

Dr. McCleese said: "Nobody spoke against that. Then someone said, 'What about a warm climate?' And then a tumultuous exchange began."

IN REVIEW

1. Explain why methane would be fragile in the Martian air.
2. According to scientists, what are the two possible sources of methane? What are the major reasons that scientists have proposed that these are the possible sources? List the major reasons some scientists are cautious about these hypotheses.
3. Explain how data from orbiting spacecraft and Earth-based telescopes are providing a stronger line of evidence for methane in the Martian atmosphere.
4. What are the implications of finding spotty regions with higher concentrations of atmospheric methane?
5. How does the presence of methane help determine the history of the Martian climate? Explain why there is great uncertainty whether Mars may have once been a much warmer planet.

Of all the bodies in the solar system, scientists have long known that the atmospheric composition of Saturn's moon Titan has been the most like Earth's. Titan, the second-largest moon in the solar system, is not much smaller then the planet Mars, and is larger than both Mercury and Pluto. Its atmospheric pressure is just a bit higher than the Earth's, and, also like on our home planet, consists mainly of nitrogen. With such a similar atmosphere to the Earth's, Titan's weather and climate have long intrigued scientists. But the atmosphere is very hazy and difficult to view with earthbound telescopes. In 2004, NASA's Cassini mission entered Saturn's orbit, with one of the major mission objectives being to understand the atmosphere and surface of Titan. Using a variety of instruments aboard the spacecraft, scientists are collecting data during several flybys of Saturn's haze-shrouded moon.

Through Saturn's Haze, A First Look at Titan

By John Noble Wilford
***The New York Times,* July 13, 2004**

Dazzling as Saturn and its rings are, to scientists the most intriguing target of exploration by the Cassini spacecraft is Saturn's giant moon Titan.

At a diameter of 3,200 miles, it is larger than the planet Mercury, and is the only moon in the solar system with a substantial atmosphere.

Cassini's first attempt to penetrate that atmosphere's thick smoggy haze for a look at Titan's surface was a mixed success. The spacecraft camera, observing visible wavelengths at a distance of 200,000 miles, captured the moon as a virtually featureless orange globe.

But scientists said the spacecraft's ability to observe through a variety of color filters and a combination of wavelengths had yielded some views of Titan's surface, though last week they were debating interpretations of what they were seeing. They have yet to detect evidence for lakes of liquid methane, as hypothesized.

Using an infrared mapping spectrometer, scientists were able to cut through the haze and distinguish broad surface features in Titan's southern hemisphere. A wide circular feature toward the equator may be a crater. Darker areas, scientists said, appeared to be regions of relatively pure water ice. Lighter areas may be covered with water ice mixed with hydrocarbon contaminants.

"We have indeed seen Titan's surface with unprecedented clarity," said Dr. Dennis Matson, the mission's chief scientist.

Some of Cassini's most striking photography showed bright clouds, probably methane, floating over a region the size of Arizona near Titan's south pole.

Better viewing lies ahead. On Oct. 26, Cassini is to make its first of 45 close approaches to Titan and use radar to map its mysterious topography.

"Titan is going to be revealed piecemeal," said Dr. Jonathan I. Lunine, a planetary scientist at the University of Arizona. "We must be patient."

Many Moons

In 1997, the Cassini mother ship and its probe Huygens began the 2-billion-mile journey to Saturn. Here are major events in the next four years.

Source: NASA

2004

JUNE 30-JULY 2 Cassini enters Saturn's orbit, making a first pass by Titan, its largest moon.

OCT. 26 First close flyby of Titan.

DEC. 24 Huygens separates and heads to Titan; later, Cassini maneuvers to avoid entering Titan's atmosphere.

2005

JAN. 14 Huygens begins descent into Titan's atmosphere.

AUG. 1 Mimas flyby.

Moon flybys, autumn 2005:
SEPT. 23 Tethys
SEPT. 25 Hyperion
OCT. 10 Dione
NOV. 25 Rhea

2006

2007

Cassini will make at least 76 orbits around Saturn during the mission and 45 near flybys of Titan, getting as close as 590 miles from the moon's surface.

2008

DEC. 3 Epimetheus flyby.

MAY 28 Last Titan flyby.

JUNE 30 End of prime mission; extensions may follow.

The New York Times

IN REVIEW

1. Explain why the first attempt to understand the atmosphere and surface of Titan was a mixed success.
2. Why did the scientists use an instrument that detects infrared light to penetrate Titan's haze rather than an instrument that detects visible light? Scientists also used color filters. How would these filters also help us to see the surface better?
3. Why were methane clouds, rather than clouds of water droplets, detected over Titan's surface? Also, why would there potentially be liquid methane on Titan's surface?
4. In the article, Dr. Lunine states that Titan is going to be revealed piecemeal. Explain what he means by this.
5. Have more recent results, including those from the Huygens probe, confirmed this early picture of Titan?

In December 2004, a probe the size of a small kitchen table separated from the Cassini spacecraft orbiting Saturn. The destination of the European Space Agency–built probe, called Huygens, was Saturn's largest moon Titan, a world shrouded in a cloudy atmospheric haze with a mysterious surface. As Huygens descended through Titan's atmosphere, it collected images and data all the way to the planet's surface, revealing an environment that is shaped by weather processes similar to Earth's. In particular, the images and data from Huygens revealed that it rains on Titan—not water rain, but liquid methane rain.

Scientists See Signs of Methane Forming as Rain on Saturn Moon

By Warren E. Leary
***The New York Times*, January 22, 2005**

Many of the weather processes found on Earth—rain falling on hills and flowing down channels into riverbeds and around islands—also are happening on Saturn's icy moon Titan, but with different materials, scientists said yesterday.

A week after a European space probe penetrated Titan's haze and landed on its surface, scientists say data show that the moon has a dynamic, eroding surface transformed by liquid methane playing the role that water serves on the Earth.

The methane—natural gas held in liquid form by the intense pressure and minus-290-degree temperatures of Titan's surface—rains from the sky and courses down highlands through channels into lakebeds and broad deltas, they said, similar to processes that take place on Earth.

"We now have the key to understanding what shapes Titan's landscape," Dr. Martin G. Tomasko of the University of Arizona said in Paris to describe the mission's first scientific results. "The physical processes shaping Titan are much the same as those shaping Earth."

Scientists said analyzing data from the Huygens probe that landed on Jan. 14 will keep them busy for years. But an early look at this information, including some 350 pictures, confirms that the moon could be a model for processes that happened on Earth billions of years ago.

"We are really extremely excited about these results," said Dr. Jean-Pierre Lebreton, mission manager for the European Space Agency's Huygens project. Scientists should start dreaming about one day sending a rover spacecraft to Titan to explore its unique terrain, he said.

The probe, carried to Saturn by NASA's Cassini spacecraft, landed on a surface with the consistency of loose sand that scientists said appeared to be moistened by methane rain. Heat generated by the spacecraft warmed the surface material, causing bursts of methane gas to boil up around the craft, scientists said.

Dr. Tomasko said the craft landed in a spot that was dry at the time, but appeared to have been rained on recently. "Does that mean yesterday or the day before, the week before? We don't really know," he said. "But the feeling is, in the place we landed, it must rain fairly frequently."

Since no liquid methane was spotted on the surface, he said, it is also possible that Huygens landed in an area that is relatively arid. "It's more like Arizona or someplace like that, where the riverbeds are dry most of the time," he went on. "But after rain, you might have open flowing liquids and pools. These pools gradually dry out; the liquid sinks down into the surface."

Dr. Tobias C. Owen of the University of Hawaii's Institute for Astronomy, noted that nitrogen was the dominant gas in Titan's upper regions but the concentrations of methane increased sharply near the surface. This indicates that there is a big source of methane at ground level, he said, most likely as a liquid very near the surface or on top that is continually renewing the supply of the gas into the atmosphere.

Scientists said the surface was covered with dark deposits from organic materials making up the haze that shrouds Titan. The material settles out of the atmosphere and when washed from higher elevations by methane rain, concentrates at the bottom of channels and riverbeds.

Instrument scans indicate that some of the surface is hard-water ice that flowing methane carves out into channels. Surface images show small rounded pebbles in a dry riverbed nearby, scientists said, adding that spectral analysis of the pebbles indicated that they were made of dirty water ice that had somehow been worn smooth by erosion.

IN REVIEW

1. List the major strands of evidence that lead scientists to believe that it rains liquid methane on Titan.
2. Explain why would it rain liquid methane on Titan rather than liquid water.
3. Describe the similarity in processes that shape both Titan's and Earth's surfaces.
4. Methane on Earth is commonly called natural gas and is very flammable. What properties of Titan's atmosphere and surface environment make it unlikely that the methane there would burst into flame?
5. The Huygens probe transmitted data from the surface of Titan for only a couple of hours. In your opinion, would it be worth sending a rover to Titan that would collect data for a much longer time? Explain.

In an interesting cycle, scientists, engineers, and science-fiction writers influence each other in ways that both lead to a deeper understanding of the universe and expand our imagination beyond our current knowledge level. For example, the theories of Albert Einstein and Stephen Hawking have been used by science-fiction writers to explain how characters travel across extreme distances of space and time. Conversely, groups of scientists and engineers are researching the possibility of greater-than-light-speed travel using ideas proposed by the *Star Trek* and *Star Wars* series. One such scientist who has been strongly linked to science fiction is Lawrence Krauss, an internationally known theoretical physicist who wrote the best-selling books *The Physics of Star Trek* and *Beyond Star Trek.* In this article, Krauss discusses how the Huygens probe landing on Titan has sparked his imagination and his motivation to be a space exploration enthusiast.

Space Probe Makes Science Fiction Wonders of Childhood Real

By Lawrence M. Krauss
***The New York Times,* January 25, 2005**

A small probe stranded on a far-away and hostile world operates for two precious hours at a temperature of 300 degrees below zero Fahrenheit, desperately transmitting information to its mother ship before that spacecraft disappears below the horizon, leaving the small explorer alone on the spongy ground of its new alien home, slowly losing power and slated to eternally rest on a frozen moon 750 million miles from Earth.

I could be accused of anthropomorphizing, but the plight of the small Cassini-Huygens probe resting by a hydrocarbon-coated ice and methane plain on Saturn's largest moon, Titan, captured my imagination far more than anything the astronauts in the International Space Station might be doing now.

What really did it for me was the orange sky. It showed with striking clarity that the science fiction wonders that I dreamed of as a child are being revealed by our unmanned space probes in a way that is both more enthralling and informative than anything likely to come from spending all of NASA's funds on a few more astronauts on the Moon, or, eventually, Mars.

I admit to having already been hooked on Internet images like those from Martian Rovers on a planet that that looks suspiciously like a smoggy sunset seen from Los Angeles. But until now, the worlds that were stunningly brought to my desktop were closer to what I might see exploring an earthly desert than to those exotic places that had so captured my imagination as a child reading science fiction stories, or looking at artists' renderings of imaginary planetary surfaces.

But there, as I clicked on the Cassini-Huygens probe Web site, the dark pebbles of dirty hydrocarbon-coated ice on the surface of Titan jumped out through an orange glow of an atmosphere unlike anything I had ever seen.

I was instead reminded of old science fiction stories. On the Web I found a recent example of the kind of thing I used to savor. This was an award-winning short story, "Slow Life" by Michael Swanwick, about human explorers seeking life on Titan.

"People talked a lot about the 'murky orange atmosphere' of Titan, but your eyes adjusted. Turn up the gain on your helmet, and the white mountains of ice were dazzling! The methane streams carved cryptic runes into the heights. Then, at the tholin-line, white turned to a rich palette of oranges, reds and yellows."

So the water-ice is dirtier and the surface darker. But the landscape of Titan is eerily similar to the one Mr. Swanwick imagined so vividly. Except that the truth is even stranger and more entrancing than his fiction.

I learned from a news conference carried out on Friday by the Cassini-Hugyens probe science team that there is evidence of active volcanoes on Titan's surface based on argon 40 in the atmosphere. But these do not spew molten lava. Instead, like the ones I concocted with my childhood chemistry set, these release flumes of water and ammonia.

There are indeed clouds and methane and hydrocarbon rainstorms, but the reality of a turbulent atmosphere of methane winds was brought home to me in a way that no writing could. With brilliant foresight, the Huygens science includes a microphone on the probe. As it fell through the clouds, beginning about 100 miles above the surface, I could listen as well as see the approaching surface as the craft sent out a stream of photos during its descent. Sitting at my computer in the middle of the night, listening to gusts of alien winds on a remote moon of Saturn was both eerie and moving.

I consider myself fortunate to be living at a time when humans are as close as they may ever come to seeing such a truly alien world with methane slush and new colors in the sky. That is probably what drew me to science in the first place. While literature has the power to lift us from the tedium of everyday existence, science at its best has the power to transport us to totally different worlds, both literal and metaphorical, to take us where our imaginations may never have otherwise traveled.

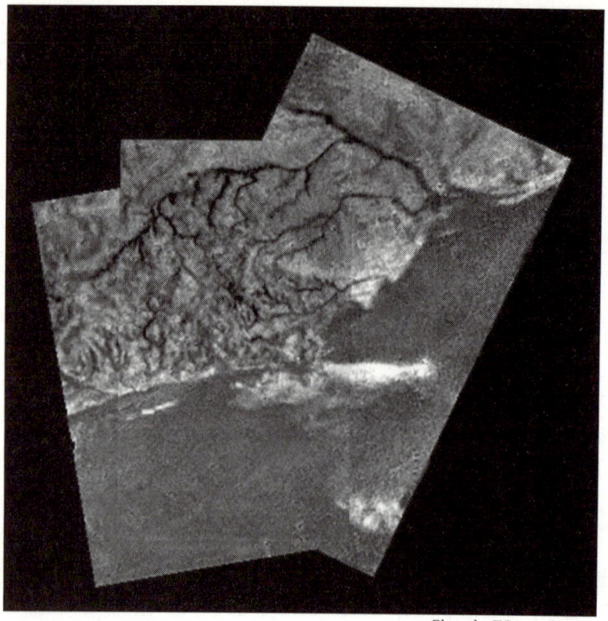

Photo by ESA via Reuters

Images taken by the Cassini-Huygens space probe show details from Saturn's largest moon, Titan, including dark pebbles of dirty hydrocarbon-coated ice.

In two short hours, one small unmanned probe changed my direct experience of our solar system in ways that I never imagined. Now I am craving for more such highs. Perhaps I will witness further probes that may dive into distant alien seas underneath frozen moons. Perhaps one will send home clear evidence of alien life existing or extinct.

Realistically, however, the future is likely to be one of cutbacks and shortfalls, with billions of dollars headed to protect populations that we put in jeopardy, or build costly missile defenses against nonexistent threats. One can only hope that there is enough imagination left in government to allow us to keep supporting the missions that do the science that can really change the way we think about our place in the universe.

To boldly go where no one has gone before in ways that only unmanned spacecraft can do will cost so little in comparison that such an effort shouldn't interfere with the current priority of allowing astronauts to have new adventures on the Moon.

It is significant in this regard that the Huygens probe was a product of the European Space Agency, working in concert with NASA and the Jet Propulsion Laboratory. This not only demonstrates that Europe is now a leading player in space exploration, but it shows that for grand human projects, like the exploration of our universe or the exploration of space and time on fundamental scales, we can and need to work together on a global scale.

This is one of the side benefits of the scientific enterprise. But even more than this, the universe continues to surprise us in ways we can never anticipate. Ultimately it is far more interesting than anything that science fiction writers or artists may imagine. Life may imitate art, but ultimately it transcends it. Which is why we sometimes need to turn to the universe itself for inspiration.

IN REVIEW

1. Why would the Huygens probe capture Lawrence Krauss's imagination more than would astronauts on the International Space Station?
2. List the features of Titan that are unlike anything that Krauss has seen before. Explain how these features are linked to his memories of science-fiction stories.
3. People often say that a picture is worth a thousand words. How have the images of Titan's surface and the sounds from the probe's microphone told a story to Lawrence Krauss?
4. Sending robotic missions to planets and moons in the solar system is much more expensive than sending human explorers. What are the benefits and drawbacks of sending humans to interesting places in the solar system compared to sending robots?
5. How has international cooperation enabled the exploration of Titan? In your opinion, should the U.S. continue to partner with foreign countries in space exploration? Explain.

In addition to being the largest planet in the solar system, Jupiter is also the planet with the strongest magnetic field. Its magnetic field is about 20,000 times stronger than Earth's, even though the planet's volume is only about 1,000 times greater than Earth's. Jupiter's magnetic field is so strong that it constantly blows away the atmosphere of its nearby moon (Io) that is being created by volcanic outgassing. Interestingly, Saturn, which is not that much smaller than Jupiter, has a much weaker magnetic field because of the low amount of metallic hydrogen at its core. However, recent data from the Cassini spacecraft, which is orbiting Saturn, show a magnetic field embedded in Saturn's beautiful rings, a totally unexpected result from the newest explorer of the outer solar system.

Spacecraft Finds Belt Around Saturn Unlike Any Ever Seen Before

By John Noble Wilford
***The New York Times,* August 6, 2004**

In its first month orbiting Saturn, the Cassini spacecraft has detected a new radiation belt in an unexpected place, its invisible swarm of trapped high-energy particles circling the planet inside the inner edge of Saturn's signature disk of luminous rings.

Scientists who reported the discovery said yesterday that they were surprised to find a relatively small radiation belt completely isolated from a planet's main radiation belts, which lie at much greater distances. They had not expected that such a concentration of energetic particles could be sustained inside the famous Saturnian rings.

The phenomenon of a single isolated belt, the scientists said, has never been observed at any other planet in the solar system. By contrast, the Van Allen belts of Earth consist of two related regions of intense radiation trapped by the planet's magnetic field.

In a NASA conference call with reporters, Dr. Donald G. Mitchell, a Cassini mission scientist from Johns Hopkins University, said the newly discovered radiation belt was detected as the spacecraft made its closest approach to Saturn immediately after rocketing into its orbit on July 1. No previous spacecraft visiting Saturn had been in the proper position to make such a discovery.

Dr. Mitchell said it was the spacecraft's imaging magnetometer that had observed the radiation belt, which extended around Saturn from about 15,000 miles above the planet's cloud tops to the inner edge of the innermost of its spectacular rings. The belt is much smaller, and the energies of its particles are less intense, than Saturn's main radiation belts.

With its discovery, Dr. Mitchell said, "we have seen something that we did not expect: that radiation belt particles can hop over obstructions like Saturn's rings."

Radiation Inside the Rings

An inner belt is unique among the planets of the solar system.

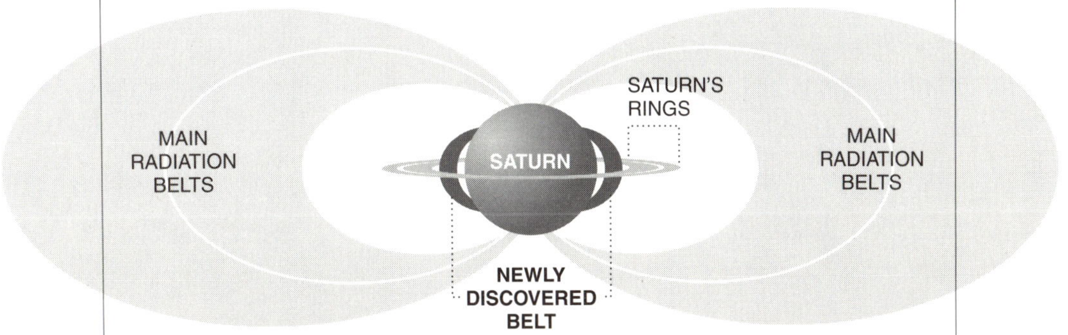

Source: NASA Jet Propulsion Laboratory

The New York Times

Other Cassini mission scientists reported puzzling patterns of lightning and thunderstorms on Saturn and observations of a striking glow emanating day and night from the planet's largest moon, Titan.

Dr. Bill Kurth, a physicist at the University of Iowa who analyzed data from the spacecraft's radio and plasma wave instrument, described the "crackle and pop" of the Saturnian electric storm as similar to what one hears "on an AM radio broadcast during a thunderstorm."

Dr. Kurth said the Saturnian storms appeared to be quite different from those observed more than 20 years ago by the two Voyager spacecraft. The storms now seem more sporadic, not as regular in timing and duration. The difference, he said, may be related to the rings' casting deep shadows over parts of the planet different from those in shadow when the Voyagers passed by.

Dr. Kevin Baines, a member of the visual and infrared imaging team at the Jet Propulsion Laboratory in Pasadena, Calif., said a glow of methane, illuminated by the Sun, revealed that Titan's atmosphere extends more than 435 miles above the surface. A small spacecraft launched from Cassini is to penetrate that atmosphere in January for the first direct investigation of the haze-shrouded Titan.

IN REVIEW

1. How do magnetic fields create radiation belts around planets?
2. Explain why the discovery of this new radiation belt was unexpected. What instrument does Cassini have that enabled it to detect the radiation belt?
3. How much did luck factor into the discovery of the radiation belt? For many discoveries, luck has come into play. How do scientists set themselves up to be lucky?
4. Ever since Voyager visited Saturn in 1981, scientists have known that there is lightning on the planet. With Cassini, scientists have seen reduced lightning strikes because of the rings' shadows on the planet. What is their hypothesis for the differerence?

For first-time telescope users, the target of choice is often Saturn. With its magnificent rings, Saturn is unlike anything that we can see with the naked eye. Larger earthbound telescopes reveal several distinctive rings, and with the Voyager spacecraft scientists discovered a complicated system of rings and moons in an intricate dance. As scientists looked at Voyager's images they began to create many theoretical models of how Saturn's ring system works, but without the data to support their models, great uncertainties remained. With the Cassini spacecraft now orbiting Saturn, scientists have a platform with several sophisticated scientific instruments that will help untangle the mysteries of Saturn and validate their theoretical ring models.

Decoding the Dance of Saturn's Rings

By Kenneth Chang
***The New York Times,* July 6, 2004**

From far away, Saturn's rings look like a single rigid disk. With more powerful telescopes, 17th-century astronomers were able to discern not a disk but a series of concentric rings.

Even closer, as when NASA's Cassini spacecraft rocketed into Saturn's orbit last week, it becomes clear that the rings are not rigid, either. While Cassini is not sharp-eyed enough to pick out the individual specks and boulders that form the rings, its photographs showed waves disturbing the rings, like ripples radiating from a pebble dropped in a pond. Here, the "pebbles" producing the ripples are Saturn's moons.

For one type of ripple, called density waves, the gravity of a moon outside the ring pulls particles outward. Certain parts of the ring, said to be in resonance with the moon, receive repeated and reinforcing outward kicks.

"Like pushing a kid on a swing," said Dr. Larry W. Esposito, a professor of astrophysical and planetary sciences at the University of Colorado and a member of the Cassini science team. "That changes this uniform density of particles in the ring. The particles clump up."

The dense clump has stronger gravity and draws other particles toward it. That results in a wave slowly spreading outward through the ring. "At about a walking pace," Dr. Esposito said.

The photographs that show the density waves were taken from the back side of the rings, the side facing away from the Sun. The dark bands are not empty regions but rather the denser clumps, which block sunlight. The light bands are the sparser regions where sunlight can pass through.

Because the orbits of many of Saturn's moons are tilted to the rings, the moons' gravity also pulls the rings up and down, creating a second type of ripples called bending waves. Just as shaking out a bedsheet creates waves fluttering down the sheet, the up and down forces create spiraling warps in the rings.

The density waves and the bending waves were not a surprise. Voyager 2 first observed them in August 1981.

Photo by NASA/Jet Propulsion Lab

Cassini postcards: moon-caused ripples and a mysterious striated ring edge.

Rather, the surprise is how well Cassini's photographs appear to fit the theoretical models that scientists have created in the last two decades.

But not everything. A portion of at least one ring appeared less like orderly waves but rather like disheveled straw, probably reflecting a turbulent flow of particles. "Don't have the foggiest," Dr. Esposito said. "Never seen anything like that before."

IN REVIEW

1. Explain how the appearance of Saturn's rings differs when viewed with the naked eye, with small telescopes, with large telescopes, from flyby spacecraft, and from orbiting spacecraft.
2. Describe the types of waves found in Saturn's rings. What causes these waves?
3. Explain how the Cassini spacecraft was positioned to view the waves. What is the importance of position in viewing the rings, and how does a spacecraft provide a unique look?
4. Describe how this article illustrates the hallmarks of science in action. What was confirmed by the data collected by the spacecraft, and what questions remain unanswered?

The discovery of the Chicxulub crater in Mexico and other strong lines of evidence have led scientists to conclude that an impact of a large asteroid or comet probably caused the extinction of the dinosaurs, as well as 75% of all the species living at that time, about 65 million years ago. Indeed, the extinction of the dinosaurs caused by the impact left mammals as a prominent species, and this paved the way for the eventual evolution of humans. While we as humans should be grateful, a potential for a repeat mass extinction is of some concern. Although scientists are not completely sure, the frequency of large impacts causing mass extinctions is about once every 50–100 million years, putting us in the range for another such event to occur. While the overall risk is small, the potential consequences are enormous. Scientists' concerns have led to such efforts as NASA's Near Earth Objects Program, which keeps an eye out for comets and asteroids that may cause problems on our home planet.

When to Post "Watch for Falling Asteroid" Sign

By Kenneth Chang
***The New York Times,* February 26, 2004**

For nine hours last month, a small band of astronomers got a scare. Their calculations indicated that a newly discovered asteroid was on a collision course with Earth and that it had a one-in-four chance of striking the planet, somewhere in the Northern Hemisphere, within days.

It did not happen, of course. Further observations in the wee hours of Jan. 14 put the asteroid in a different orbit, at no threat to Earth.

But the episode, the latest in a series of false alarms, pointed up the disquieting prospect of an asteroid's showing up suddenly on Earth's doorstep with no time for Hollywood heroics. No guidelines exist for who should have been informed when, if the threat had been real, and what emergency measures should have been taken. At least one planetary scientist says he was ready to recommend a public warning.

"I would not have been comfortable with being quiet through the next morning," said the scientist, Dr. Clark R. Chapman of the Southwest Research Institute in Boulder, Colo., who was involved in discussions of the meteor that night. "I think the public should be informed of that high a probability of that big an event occurring."

Dr. Chapman presented a paper on Monday at the Planetary Defense Conference in Garden Grove, Calif., recapping the sequence of events from the evening of Jan. 13 through the morning of Jan. 14.

The asteroid, now designated 2004 AS1, was not a planet killer, like the one that probably caused the extinction of the dinosaurs 65 million years ago, or even a city killer. Believed at the time to be about 100 feet across, 2004 AS1 would most likely have exploded with the force of a one-megaton bomb several miles up in the atmosphere. The shock waves would have set off hurricane-force winds.

"It's right at that boundary line, so we don't know how much damage it would have done," Dr. Chapman said. An object half the size would explode harmlessly. One twice as wide would be catastrophic, he said.

Dr. Donald Yeomans, head of the Near Earth Objects Program Office of NASA, said he would not have raised an alarm until a second set of observations confirmed the collision path. He said he hoped the episode would prompt guidelines for how warnings should be handled. "Hopefully, policy makers will take the ball and run with it," Dr. Yeomans said.

For several years, NASA has spent $3.5 million a year on surveys to locate and map the asteroids zooming through Earth's neighborhood, but the program, called Spaceguard, focuses on the larger asteroids, and the presumption that a potential impact would be years away.

"The Spaceguard Survey is not designed to detect objects on their final approach to Earth," said Dr. David Morrison, a NASA scientist.

Last month's episode put a spotlight on a dedicated but largely ad hoc network of professional and amateur astronomers. On Jan. 13, the Minor Planet Center at the Harvard-Smithsonian Center for Astrophysics received the latest observations from Linear, short for Lincoln Near-Earth Asteroid Research, an automated telescope survey in New Mexico that searches for new asteroids.

Dr. Timothy Spahr, an astronomer at the Minor Planet Center, sifted through, finding several interesting ones. He ran the computer programs that calculated possible orbits for the asteroids and posted them on a Web site so that other astronomers could confirm the locations. Dr. Spahr then went to dinner with a colleague visiting from Hawaii. He did not notice that the orbit he had calculated for one of the new asteroids went straight through Earth.

A short time later, an amateur astronomer in Germany noticed on the Minor Planet Center Web site the odd prediction that 2004 AS1 was to brighten by a factor of 40 in the next day, an indication like oncoming headlights that the object was on the express route to Earth. Soon, both amateur and professional astronomers were buzzing about this object on an Internet bulletin board devoted to asteroids.

Dr. Steven R. Chesley, a senior engineer working in the Near Earth Objects office at the Jet Propulsion Laboratory in Pasadena, Calif., called the Minor Planet Center asking for more details about the asteroid. Dr. Spahr was still out to dinner. Dr. Brian Marsden, director of the center, was working late and sent the details of the Linear observations—four images taken over a one-hour period—to Dr. Chesley.

The pictures of the asteroid show only its position in its sky; how far away it is, whether a million miles or 50 million, is necessarily a guess based on sparse data. Dr. Marsden then changed the information on the center's Web site, now saying that the asteroid was actually moving away from Earth—an orbit that still fit the observations.

Dr. Spahr did not learn of the worries until he arrived home at 10:30 p.m. "We never ever imagined that would happen," he said. "I actually felt pretty silly."

He and Dr. Chesley exchanged information and calculated. Dr. Chesley's calculations of possible orbits put the chances of impact between 10 percent and 40 percent.

What the astronomers needed was a new spotting of the asteroid to see how much it had moved since the Linear observations. But with skies cloudy over most of Europe and North America, asteroid experts scrambled to find amateur sky watchers with a clear view.

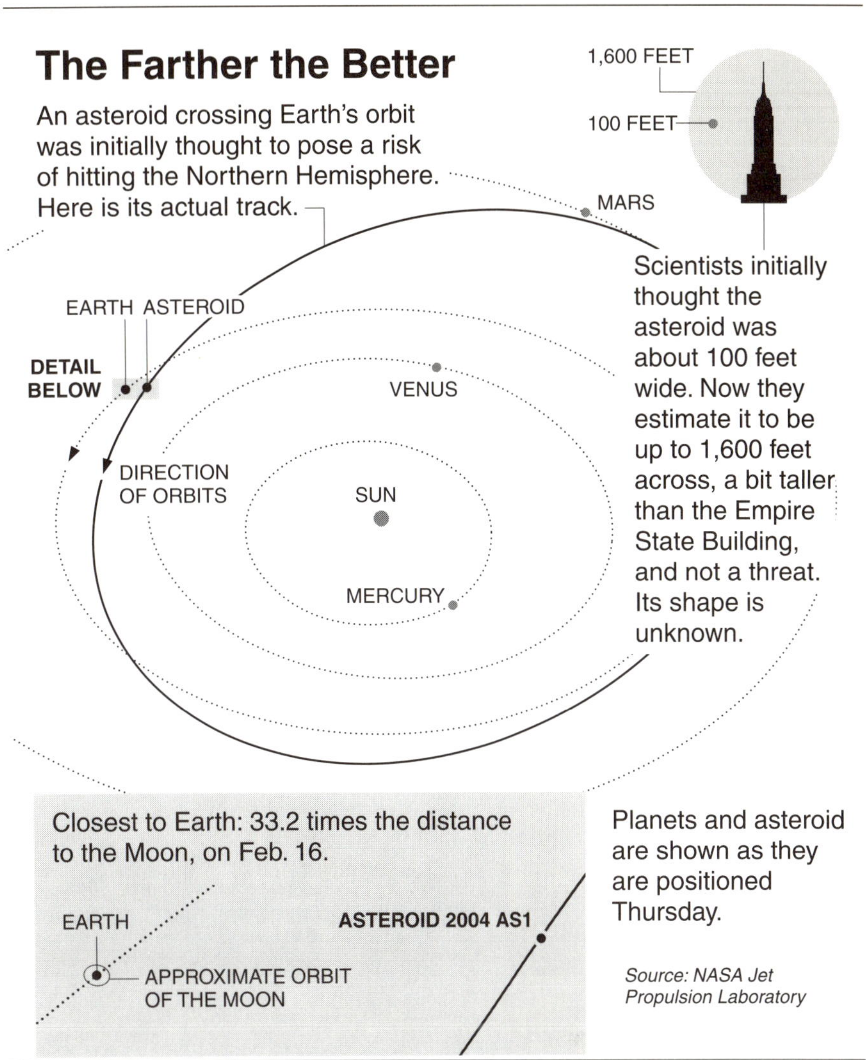

The New York Times

Dr. Spahr determined where in the sky the asteroid would be if it were indeed on the collision course. At 3:30 a.m. Eastern time, an amateur astronomer in Colorado looked in that swath of the sky, and saw nothing. The next night, Linear spotted 2004 AS1 again, showing that it was much farther out and not on a collision course at all. (Because it is farther away, its size is now estimated to be 1,600 feet wide.) The errors in the data from the previous night had been somewhat larger than usual, and that slight deviation in location had led Dr. Spahr's software to put the asteroid on the alarming path.

Later observations showed that 2004 AS1, which came no closer to Earth than eight million miles, has no chance of hitting this century.

IN REVIEW

1. List the uncertainties that scientists had about the asteroid that came close to Earth's position.
2. How can a network of automated telescopes help in detecting asteroids that may impact Earth?
3. Describe the major roles that amateur astronomers play in observing asteroids and comets.
4. How did scientists determine that the asteroid did not pose a threat to the Earth?
5. In your opinion, is the world adequately prepared for a potentially major impact? Explain. Also, how much money and time should be spent in searching for near Earth asteroids and comets? Provide reasoning for your estimate.

Comets have fascinated sky watchers throughout history. With their brilliant and beautiful tails, comets have been used as omens of both good and bad luck, in some cases influencing the decisions of important leaders throughout history. Today, scientists examine comets to understand the beginnings of our solar system. From telescope observations, astronomers have learned that comets contain water and carbon dioxide, as well as complex organic molecules. Early comet impacts may have seeded our young Earth with both water and organics, providing the needed ingredients for life. To help us understand more about comets and their potential importance, NASA's Stardust mission has provided a close-up look of the comet Wild 2, and it will bring back material captured from the comet's to Earth.

To Surprise of Researchers, Comet Has a Personality

By John Noble Wilford
***The New York Times,* June 18, 2004**

The most revealing close-up pictures ever taken of a comet have scientists shaking their heads in astonishment. The rugged, diverse landscape of the comet Wild 2 is unlike anything they have ever seen or imagined: towering columns and spires rising above steep-walled craters and violent jets of gas and dust shooting skyward.

"We were totally stunned by what we saw," Dr. Donald Brownlee of the University of Washington said yesterday at a NASA news conference announcing findings from the Stardust spacecraft, which flew within 147 miles of the comet on Jan. 2.

Wild 2, named after a Swiss astronomer and pronounced "vilt," also seemed to make a strong impression on the 770-pound spacecraft. Blasts of particles rising like antiaircraft flak rocked the Stardust. A dozen particles, some larger than a bullet, penetrated the outer layer of its protective shielding.

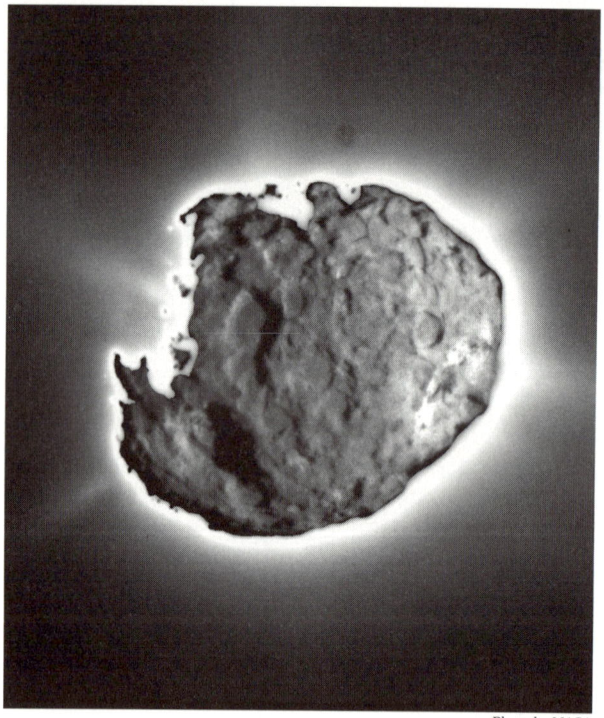

Photo by NASA

A photograph taken in January by the Stardust spacecraft just 147 miles from the Wild 2 comet shows the comet pocked with craters.

"These things were like a thunderbolt," Dr. Anthony Tuzzolino, a senior scientist at the Fermi Institute at the University of Chicago, said in a university statement. "I didn't anticipate running into this kind of show."

But the spacecraft survived and is returning cometary dust samples to Earth. Dr. Brownlee, the mission's principal investigator, and other researchers are making their first detailed reports of the encounter in articles today in the journal Science.

The findings left the scientists perplexed. The comet, three miles in diameter, appeared to be unlike any previous object examined in the solar system.

Was it typical of comets or an oddball? Astronomers had long supposed that comets resembled dirty snowballs, loose conglomerations of ice and rubble like Halley's Comet, and perhaps released some dust and gas, but not with the force encountered by the spacecraft.

"We are astonished and intrigued by seeing what we saw with the Stardust encounter," said Dr. Claudia Alexander, an astronomer at NASA's Jet Propulsion Laboratory in Pasadena, Calif., who specializes in comet research. The televised news conference was held at the laboratory, which directed the mission.

Dr. Brownlee said Wild 2's landscape reminded him of the spires and buttes of Monument Valley in Arizona, left standing after erosion wiped away the surrounding terrain. The comet's spires rise more than 300 feet, and its craters are as much as 500 feet wide. If these were indeed erosion features, he said, then the comet has probably lost more than 300 feet of its surface layer as it has traveled through the solar system.

Perhaps the strangest surface structures, researchers said, were two deep depressions with flat floors and steep walls. They looked like two giant footprints. On a map of the comet, they are labeled Left Foot and Right Foot.

Scientists theorized that evaporation and hits from objects created the comet's rugged surface. The violent jets may form when the sun shines on icy areas near or just below the surface. The ice becomes a gas, blasting into the vacuum of space at hundreds of miles an hour.

Dr. Alexander said scientists were struck by the starkly different appearances of Wild 2 and Phoebe, a small moon of Saturn that the Cassini spacecraft flew by last week. Phoebe is thought to be a captured object that originated, like Wild 2 and other comets, on the cold fringes of the solar system beyond the planets.

The two objects may well be from the same distant region, Dr. Alexander said, but have different histories. Phoebe was captured long ago by Saturn and remained in its orbit. But Wild 2 has presumably undergone much different environments as it traveled far out in space and then closer to the heat of the Sun.

Dr. Benton Clark, a Stardust scientist with Lockheed Martin Space Systems in Denver, reported that spacecraft instruments found the comet's dust particles to be rich in organic material. This could be significant, he said, because some astronomers have contended that comets might have seeded Earth with many of the organic building blocks for life.

Flying through the dust around Wild 2, the spacecraft gathered thousands of particles that are now being returned to Earth for closer investigation. A capsule holding the exotic cargo is to make a soft landing in the Utah desert in January 2006.

IN REVIEW

1. Describe the features that stunned scientists when they looked at close-up images of comet Wild 2.
2. What hazards did the Stardust spacecraft encounter as it traveled through the comet's tail? How did engineers design the spacecraft to withstand these hazards?
3. What are the explanations scientists are using to describe the presence of the tall spires and deep craters on the comet? How do they explain the violent jets of gas being emitted from the comet?
4. List the similarities and differences between comet Wild 2 and Saturn's moon Phoebe. Why would we expect these objects to be similar, and what might explain some of their differences?
5. What is the significance of the organic materials remotely sensed by the spacecraft's instruments? How might scientists learn more about the comet, including its organic materials, by carefully examining samples of the comet's tail that will be returned to Earth?

Scientists have long postulated that comets originate from vast reservoirs in the outer solar system. Many believe that these reservoirs are the remnants of the original dusty and gaseous disk that formed the solar system. However, recent discoveries of relatively large icy objects that orbit at extreme distances have made some scientists rethink their hypotheses. The recent observations of the Hubble and Spitzer space telescopes showing a large number of stars being born in close quarters have compelled astronomers to investigate computer models in order to understand how large objects such as comets may be swapped among planetary systems as they form in these stellar nurseries. If such swapping occurs, the solar system may be a much more interesting mix of materials than scientists earlier thought.

Sun Might Have Exchanged Hangers-On with Rival Star

By Dennis Overbye
***The New York Times*, December 2, 2004**

The Sun may have captured thousands or even millions of asteroids from another planetary system during an encounter more than four billion years ago, astronomers are reporting today.

Such an interstellar ballet would explain many mysteries of the outer solar system—including the strange behavior of the recently discovered Sedna, the system's most distant known object, which occupies a strange elongated orbit far beyond Pluto.

The astronomers' calculations, from supercomputer simulations, suggest that gravity from the star at the center of the other planetary system could have kicked Sedna out of a more conventional orbit. In the process, the Sun and the other star would have swapped their outer entourages. Indeed, the astronomers estimate that there is a 10 percent probability that Sedna itself was one of those strangers.

If the alien asteroids could be found and studied, these bodies could provide testimony to the conditions under which the Sun and the solar system formed, a time otherwise lost in the mists. Astronomers say the Sun was born 4.5 billion years ago as part of a dense cluster of more than 1,000 stars that has long since disappeared. The star that nearly collided with our solar system so long ago could be on the other side of the galaxy by now.

"I don't think anyone has considered that extrasolar planets would be in our own solar system," said Scott J. Kenyon of the Harvard-Smithsonian Center for Astrophysics. He is the co-author, with Benjamin C. Bromley of the University of Utah, of a paper being published today in Nature.

Their paper is the latest in a series of efforts to consider an intruding star as a way of explaining the weird properties of Sedna.

"Sedna surprised the hell out of everybody," said Harold F. Levison of the Southwestern Research Institute in Boulder, Colo., who has proposed a slightly different intruder scenario, with Alessandro Morbidelli of the Observatory of the Côte d'Azur in France.

The strange planetoid was discovered last spring. About 600 to 1,000 miles in diameter, Sedna travels in an elongated oval of an orbit far outside the main planets of the solar system, taking roughly 10,500 years to complete an orbit. By comparison, distant Pluto takes about 248 years to complete a trip around the Sun. At its closest approach, Sedna, named after the Inuit goddess who dwells in the frigid depths of the seas, will still be about seven billion miles from the Sun, 75 times farther than the Earth, which circles at a comfortable 93 million miles—one astronomical unit, in cosmic lingo.

Before Sedna's discovery, astronomers knew of no objects beyond 50 astronomical units from the Sun, where a disk of icy worlds known as the Kuiper Belt, left over from the formation of Neptune and Uranus, abruptly falls off. (They have deduced the existence of a vast halo of icy objects from which comets occasionally descend, known as the Oort cloud, extending much farther out, but these objects cannot be seen.)

Out there, Sedna is far beyond the gravitational influence of the other planets, so how did it get so far out? Both Dr. Kenyon's group and Dr. Levison's agree that a kick from a passing star is the best and perhaps the only way to do it, but they disagree on the details.

Dr. Kenyon and Dr. Bromley conclude that a star and its disk of planetary building materials passing from 160 astronomical units or so from the Sun could not only lift Sedna into its present location, but also truncate the Kuiper Belt, thus explaining its abrupt edge.

Dr. Levison said his and Dr. Morbidelli's work favored a more distant and gentle kick from a distance of about 800 astronomical units. He pointed out that Dr. Kenyon's scenario predicted a plethora of objects in Sedna-like orbits but slightly closer and thus easier to see that are nonetheless not seen.

Dr. Kenyon acknowledged that this was a potential problem.

Either encounter would also leave alien planetoids in our solar system (and some of ours in the alien system) orbiting at a steep angle to the plane in which the planets go around. And so the next step is to search for such objects.

Sedna itself has only a moderately inclined orbit, the astronomers say. A more likely candidate for an extra-solar origin is another icy wanderer, known as 2000 CR105, about half the size of Sedna, discovered out beyond Neptune in 2000. Its orbit is inclined 20 degrees to the planets.

The detection of objects with inclinations of 40 degrees or more, the authors write in Nature, "would clinch the case for extrasolar objects in the solar system."

Alan P. Boss, a planetary expert at the Carnegie Institution of Washington, called the idea that Sedna could have been a captured object formed around a different star "intriguing even if it is a low-probability event."

Dr. Bromley said the study of such objects could help scientists understand how stars and solar systems form and whether the forces that shaped Sedna and the solar system are common to other systems. "We are a long way off from having the technology to observe Sedna-like objects around other stars," he wrote in an e-mail message, "but a captured planet would be within our reach."

Dr. Levison said these objects could be the few relics left with information about the star cluster that gave birth to the Sun.

"All our cousins took off and we have no idea about the size of our family," he said. "This gives us a little cousin to study."

IN REVIEW

1. What properties of the recently discovered Sedna are perplexing to scientists?
2. Using computer models, what explanations have scientists come up with to unravel Sedna's mysteries?
3. List the similarities and differences between the work done by the two scientific groups discussed in the article, Dr. Kenyon's group and Dr. Levison's group.
4. Why would objects with inclined orbits be more likely candidates to be from another solar system?
5. Why would Alan Boss call the idea that Sedna was captured from another star by our Sun "a low-probability event"? What additional data and observations are necessary to support the computer models suggesting that Sedna comes from outside our solar system?

The movement of the Earth's crust, which is broken into plates, is one of the unique features of our planet. As more than a dozen plates slowly collide and separate, mountain ranges rise, rift valleys split, and volcanic islands emerge. More sudden and violent plate movement results in earthquakes. Such was the case in December 2004, when a release of huge stresses between two crustal plates near Sumatra, Indonesia, created a violent earthquake below the ocean. The resulting tsunami caused a wave of destruction for thousands of miles around. Understanding the motion of the plates and the resulting impacts is the work of countless numbers of scientists who seek, in part, to prevent the huge loss of life associated with catastrophic tsunamis.

Deadly and Yet Necessary, Quakes Renew the Planet

By William J. Broad
***The New York Times*, January 11, 2005**

They approach the topic gingerly, wary of sounding callous, aware that the geology they admire has just caused a staggering loss of life. Even so, scientists argue that in the very long view, the global process behind great earthquakes is quite advantageous for life on earth—especially human life.

Powerful jolts like the one that sent killer waves racing across the Indian Ocean on Dec. 26 are inevitable side effects of the constant recycling of planetary crust, which produces a lush, habitable planet. Some experts refer to the regular blows—hundreds a day—as the planet's heartbeat.

The advantages began billions of years ago, when this crustal recycling made the oceans and atmosphere and formed the continents. Today, it builds mountains, enriches soils, regulates the planet's temperature, concentrates gold and other rare metals and maintains the sea's chemical balance.

Plate tectonics (after the Greek word "tekton," or builder) describes the geology. The tragic downside is that waves of quakes and volcanic eruptions along plate boundaries can devastate human populations.

"It's hard to find something uplifting about 150,000 lives being lost," said Dr. Donald J. DePaolo, a geochemist at the University of California, Berkeley. "But the type of geological process that caused the earthquake and the tsunami is an essential characteristic of the earth. As far as we know, it doesn't occur on any other planetary body and has something very directly to do with the fact that the earth is a habitable planet."

Many biologists believe that the process may have even given birth to life itself.

The main benefits of plate tectonics accumulate slowly and globally over the ages. In contrast, its local upheavals can produce regional catastrophes, as the recent Indian Ocean quake made clear.

Even so, scientists say, the Dec. 26 tsunamis may prove to be an ecological boon over the decades for coastal areas hardest hit by the giant waves.

Dr. Jelle Zeilinga de Boer, a geologist at Wesleyan University who grew up in Indonesia and has studied the archipelago, says historical evidence from earlier tsunamis suggests that the huge waves can distribute rich sediments from river systems across coastal plains, making the soil richer.

"It brings fertile soils into the lowlands," he said. "In time, a more fertile jungle will develop."

Dr. de Boer, author of recent books on earthquakes and volcanoes in human history, added that great suffering from tectonic violence was usually followed by great benefits as well. "Nature is reborn with these kinds of terrible events," he said. "There are a lot of positive aspects even when we don't see them."

Plate tectonics holds that the earth's surface is made up of a dozen or so big crustal slabs that float on a sea of melted rock. Over ages, this churning sea moves the plates as well as their superimposed continents and ocean basins, tearing them apart and rearranging them like pieces of a puzzle.

The process starts as volcanic gashes spew hot rock that spreads out across the seabed. Eventually, hundreds or thousands of miles away, the cooling slab collides with other plates and sinks beneath them, plunging back into the hot earth.

The colliding plates grind past one another about as fast as fingernails grow and over time produce mountains and swarms of earthquakes as frictional stresses build and release. Meanwhile, parts of the descending plate melt and rise to form volcanoes on land.

The recent cataclysm began in a similar manner as volcanic gashes in the western depths of the Indian Ocean belched molten rock to form the India plate. Its collision with the Burma plate created the volcanoes of Sumatra as well thousands of earthquakes, including the magnitude 9.0 killer.

A Lush Planet Emerges From Epic Violence

Earthquakes and resulting tsunamis are unfortunate byproducts of a process necessary to renew the earth. Subduction — the collision of plates that make up the earth's crust in which one plate is pulled below another — plays a critical role in recycling carbon dioxide, which is needed to regulate the planet's temperature.

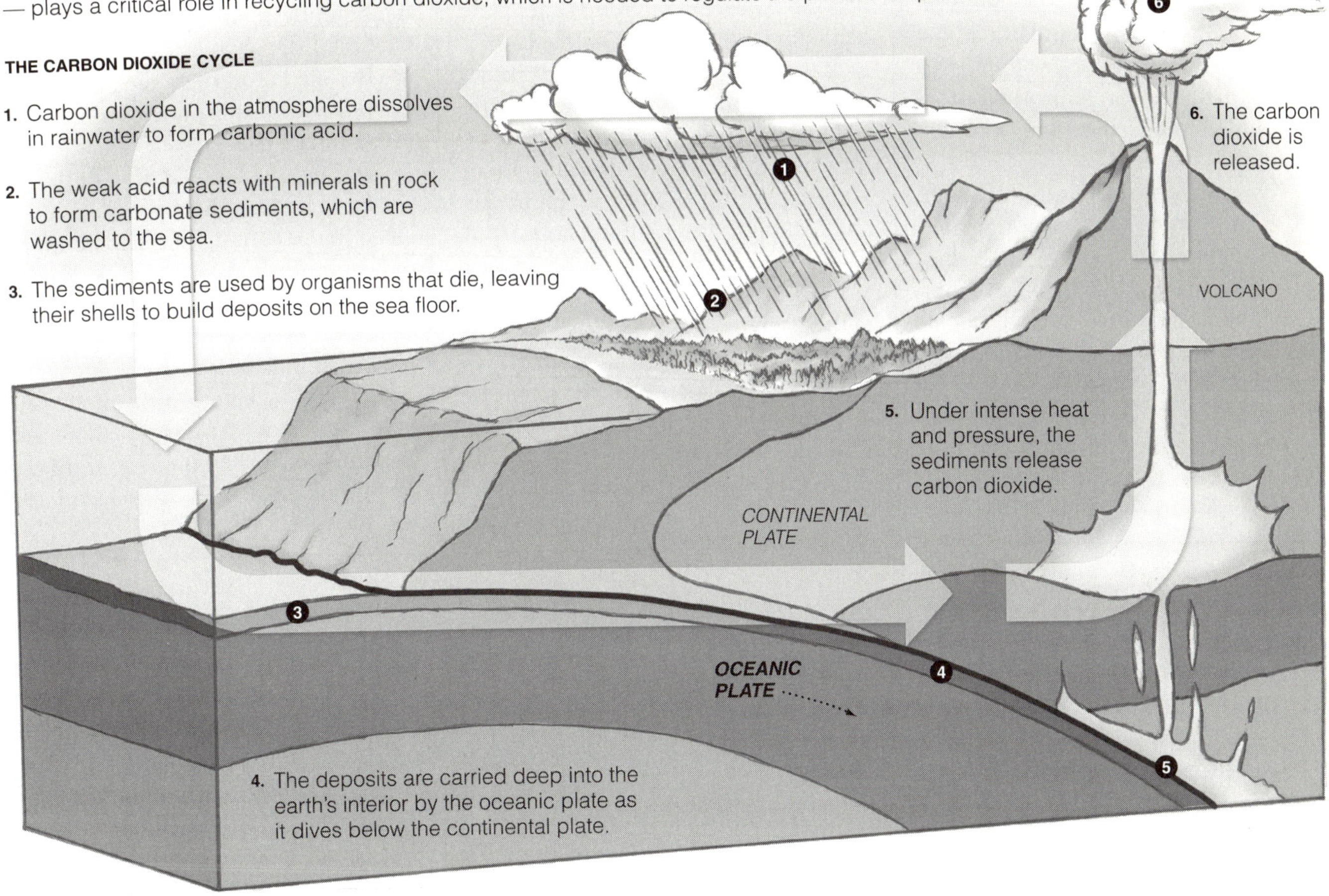

Other Benefits of Plate Tectonics

Made oceans and atmosphere Most geologists believe the earth's primordial ocean and atmosphere arose as volcanoes spewed water vapor, nitrogen, carbon dioxide and other gases.

Made land Simple volcanism made islands like Sumatra and Java, where scores of volcanoes still erupt. Additionally, geologists say plate tectonics builds continents and mountain ranges.

Improves soil Volcanic soils, some of the world's richest, are often used to form plantations for producing coffee, sugar, rubber, coconuts, palm oil, tobacco, pepper, tea and cocoa.

Begat life Many biologists believe the earth's first organisms arose in the deep sea along volcanic gashes and that microbes known as thermophiles, which thrive today in such hot regions, are their direct descendants.

Regulates global chemistry Experts say the world ocean passes through the tectonic system once every million years or so, increasing nutrients in the biosphere and regulating a host of elements and compounds.

Concentrates metals Water streaming through the seabed's hot volcanic gashes concentrates silver, gold and other metals into rich deposits that are mined after plate tectonics nudges them up onto land.

Sources: "Biogeochemistry: An Analysis of Global Change" by William H. Schlesinger; "Understanding Earth" by Frank Press, Raymond Siever, John Grotzinger and Thomas H. Jordan

David Constantine, Al Granberg/The New York Times

But despite such staggering losses of life, said Robert S. Detrick Jr., a geophysicist at the Woods Hole Oceanographic Institution, "there's no question that plate tectonics rejuvenates the planet."

Moreover, geologists say, it demonstrates the earth's uniqueness. In the decades after the discovery of plate tectonics, space probes among the 70 or so planets and moons that make up the solar system found that the process existed only on earth—as revealed by its unique mountain ranges.

In the book "Rare Earth" (Copernicus, 2000), which explored the likelihood that advanced civilizations dot the cosmos, Dr. Peter D. Ward and Dr. Donald Brownlee of the University of Washington argued in a long chapter on plate tectonics that the slow recycling of planetary crust was uncommon in the universe yet essential for the evolution of complex life.

"It maintains not just habitability but high habitability," said Dr. Ward, a paleontologist. (Dr. Brownlee is an astronomer.) Most geologists believe that the process yielded the earth's primordial ocean and atmosphere, as volcanoes spewed vast amounts of water vapor, nitrogen, carbon dioxide and other gases. Plants eventually added oxygen. Meanwhile, many biologists say, the earth's first organisms probably arose in the deep sea, along the volcanic gashes.

"On balance, it's possible that life on earth would not have originated without plate tectonics, or the atmosphere, or the oceans," said Dr. Frank Press, the lead author of "Understanding Earth" (Freeman, 2004) and a past president of the National Academy of Sciences.

The volcanoes of the recycling process make rich soil ideal for producing coffee, sugar, rubber, coconuts, palm oil, tobacco, pepper, tea and cocoa. Water streaming through gashes in the seabed concentrates copper, silver, gold and other metals into rich deposits that are often mined after plate tectonics nudges them onto dry land.

Experts say the world ocean passes through the rocky pores of the tectonic system once every million years or so, increasing nutrients in the biosphere and regulating a host of elements and compounds, including boron and calcium.

Dr. William H. Schlesinger, dean of the Nicholas School of the Environment and Earth Sciences at Duke, says one vital cycle keeps adequate amounts of carbon dioxide in the atmosphere. Though carbon dioxide is thought to cause excessive greenhouse-gas warming of the planet, an appreciable level is needed to keep the planet warm enough to support life.

"Having plate tectonics complete the cycle is absolutely essential to maintaining stable climate conditions on earth," Dr. Schlesinger said. "Otherwise, all the carbon dioxide would disappear and the planet would turn into a frozen ball."

Dr. Press, who was President Jimmy Carter's science adviser, said the challenge in the coming decades would be to keep enjoying the benefits of plate tectonics while improving our ability to curb its deadly byproducts.

"We're making progress," Dr. Press said. "We can predict volcanic explosions and erect warning systems for tsunamis. We're beginning to limit the downside effects."

IN REVIEW

1. Identify and list the different effects of plate tectonics.
2. Contrast the gradual effects of plate tectonics with those that happen rapidly.
3. Describe how plate tectonic cycles were demonstrated by the December 2004 earthquake near Sumatra.
4. How would plate tectonics make Earth highly habitable? Compare the major life-supporting benefits of the Earth's plate tectonics to a planet like Mars that does not have plate tectonics. What are the implications for finding life on Mars?
5. How will understanding plate tectonics better prevent major losses of life associated with earthquakes and tsunamis? How would deepening our knowledge through science benefit the technological advances made by engineers and entrepreneurs who are concerned with public safety?

Without its atmosphere, the Earth's average temperature would be much lower, and most, if not all, of the water on its surface would be frozen. Mars, with an atmosphere about 1% that of Earth, is much colder, and the only water on its surface is water ice. In essence, the Earth's atmosphere acts as a greenhouse to keep the planet's water liquid and increases its hospitableness toward life. Venus too has a greenhouse effect, but because the pressure of its atmosphere is about 100 times greater than Earth's, its average temperature, at over 800°F, is the greatest in the solar system. Venus's searing temperatures would boil away any liquid water on its surface; the planet is a hellish place and not able to support life. As combustion of fossil fuels increases the amount of gases that contribute to the greenhouse effect in the Earth's atmosphere, scientists worry about global climate change that may have serious impacts on Earth's life.

Big Arctic Perils Seen in Warming

By Andrew C. Revkin
***The New York Times,* October 30, 2004**

A comprehensive four-year study of warming in the Arctic shows that heat-trapping gases from tailpipes and smokestacks around the world are contributing to profound environmental changes, including sharp retreats of glaciers and sea ice, thawing of permafrost and shifts in the weather, the oceans and the atmosphere.

The study, commissioned by eight nations with Arctic territory, including the United States, says the changes are likely to harm native communities, wildlife and economic activity but also to offer some benefits, like longer growing seasons. The report is due to be released on Nov. 9, but portions were provided yesterday to The New York Times by European participants in the project.

While Arctic warming has been going on for decades and has been studied before, this is the first thorough assessment of the causes and consequences of the trend.

It was conducted by nearly 300 scientists, as well as elders from the native communities in the region, after representatives of the eight nations met in October 2000 in Barrow, Alaska, amid a growing sense of urgency about the effects of global warming on the Arctic.

The findings support the broad but politically controversial scientific consensus that global warming is caused mainly by rising atmospheric concentrations of heat-trapping greenhouse gases, and that the Arctic is the first region to feel its effects. While the report is advisory and carries no legal weight, it is likely to increase pressure on the Bush administration, which has acknowledged a possible human role in global warming but says the science is still too murky to justify mandatory reductions in greenhouse-gas emissions.

The State Department, which has reviewed the report, declined to comment on it yesterday.

The report says that "while some historical changes in climate have resulted from natural causes and variations, the strength of the trends and the patterns of change that have emerged in recent decades indicate that human influences, resulting primarily from increased emissions of carbon dioxide and other greenhouse gases, have now become the dominant factor."

The Arctic "is now experiencing some of the most rapid and severe climate change on Earth," the report says, adding, "Over the next 100 years, climate change is expected to accelerate, contributing to major physical, ecological, social and economic changes, many of which have already begun."

Scientists have long expected the Arctic to warm more rapidly than other regions, partly because as snow and ice melt, the loss of bright reflective surfaces causes the exposed land and water to absorb more of the sun's energy. Also, warming tends to build more rapidly at the surface in the Arctic because colder air from the upper atmosphere does not mix with the surface air as readily as at lower latitudes, scientists say.

The report says the effects of warming may be heightened by other factors, including overfishing, rising populations, rising levels of ultraviolet radiation from the depleted ozone layer (a condition at both poles). "The sum of these factors threatens to overwhelm the adaptive capacity of some Arctic populations and ecosystems," it says.

Prompt efforts to curb greenhouse-gas emissions could slow the pace of change, allowing communities and wildlife to adapt, the report says. But it also stresses that further warming and melting are unavoidable, given the century-long buildup of the gases, mainly carbon dioxide.

Several of the Europeans who provided parts of the report said they had done so because the Bush administration had delayed publication until after the presidential election, partly because of the political contentiousness of global warming.

But Gunnar Palsson of Iceland, chairman of the Arctic Council, the international body that commissioned the study, said yesterday that there was "no truth to the contention that any of the member states of the Arctic Council pushed the release of the report back into November." Besides the United States, the members are Canada, Denmark, Finland, Iceland, Norway, Russia and Sweden.

Mr. Palsson said all the countries had agreed to delay the release, originally scheduled for September, because of conflicts with another international meeting in Iceland.

The American scientist directing the assessment, Dr. Robert W. Corell, an oceanographer and senior fellow of the American Meteorological Society, said the timing was set during diplomatic discussions that did not involve the scientists.

He said he could not yet comment on the specific findings, but noted that the signals from the Arctic have global significance.

"The major message is that climate change is here and now in the Arctic," he said.

The report is a profusely illustrated window on a region in remarkable flux, incorporating reams of scientific data as well as observations by elders from native communities around the Arctic Circle.

The potential benefits of the changes include projected growth in marine fish stocks and improved prospects for agriculture and timber harvests in some regions, as well as expanded access to Arctic waters.

But the list of potential harms is far longer.

The retreat of sea ice, the report says, "is very likely to have devastating consequences for polar bears, ice-living seals and local people for whom these animals are a primary food source."

Oil and gas deposits on land are likely to be harder to extract as tundra thaws, limiting the frozen season when drilling convoys can traverse the otherwise spongy ground, the report says. Alaska has already seen the "tundra travel" season on the North Slope shrink to 100 days from about 200 days a year in 1970.

The report concludes that the consequences of the fast-paced Arctic warming will be global. In particular, the accelerated melting of Greenland's two-mile-high sheets of ice will cause sea levels to rise around the world.

IN REVIEW

1. What is known about greenhouse gas emissions and global warming? What is uncertain?
2. Explain why the Arctic and Alpine regions have seen the biggest impacts with global warming. What other factors may contribute to the impacts on the Arctic?
3. Why did the report conclude that further warming and Arctic melting are unavoidable?
4. List the different political effects and concerns that were discussed in the article. Speculate as to why politics has such a strong influence on the debate associated with global climate change.
5. In your opinion, should the countries of the world take strong measures to curb emissions of greenhouse gases? Explain.

Antarctica is covered by so much ice that, if it all melted, sea level would rise by more than 200 feet. Fortunately, even the most pessimistic predictions of models of global warming do not suggest that such large-scale melting could occur any time soon. Nevertheless, some substantial change seems to be afoot in Antarctica, perhaps suggesting that global warming might affect that continent and sea level more than most scientists had assumed.

Antarctica, Warming, Looks Ever More Vulnerable

By Larry Rohter
***The New York Times,* January 25, 2005**

Correction Appended

From an airplane at 500 feet, all that is visible here is a vast white emptiness. Ahead, a chalky plain stretches as far as the eye can see, the monotony broken only by a few gentle rises and the wrinkles created when new sheets of ice form.

Under the surface of that ice, though, profound and potentially troubling changes are taking place, and at a quickened pace. With temperatures climbing in parts of Antarctica in recent years, melt water seems to be penetrating deeper and deeper into ice crevices, weakening immense and seemingly impregnable formations that have developed over thousands of years.

As a result, huge glaciers in this and other remote areas of Antarctica are thinning and ice shelves the size of American states are either disintegrating or retreating—all possible indications of global warming. Scientists from the British Antarctic Survey reported in December that in some parts of the Antarctic Peninsula hundreds of miles from here, large growths of grass are appearing in places that until recently were hidden under a frozen cloak.

"The evidence is piling up; everything fits," Dr. Robert Thomas, a glaciologist from NASA who is the lead author of a recent paper on accelerating sea-level rise, said as the Chilean Navy plane flew over the sea ice here on an unusually clear day late in November. "Around the Amundsen Sea, we have surveyed a half dozen glaciers. All are thinning, in some cases quite rapidly, and in each case, the ice shelf is also thinning."

Photo by British Antarctic Survey via Associated Press

The Larsen B ice shelf collapsed over a 35-day period early in 2002, losing more than a quarter of its total mass.

The relationship between glaciers (essentially frozen rivers) and ice shelves (thick plates of ice protruding from the land and floating on the ocean) is complicated and not fully understood. But scientists like to compare the spot where the "tongue" of a glacier flows to sea in the form of an ice shelf to a cork in a bottle. When the ice shelf breaks up, this can allow the inland ice to accelerate its march to the sea.

"By themselves, the tongue of the glacier or the cork in the bottle do not represent that much," said Dr. Claudio Teitelboim, the director of the Center for Scientific Studies, a private Chilean institution that is the partner of the National Aeronautics and Space Administration in surveying the ice fields of Antarctica and Patagonia. "But once the cork is dislodged, the contents of the bottle flow out, and that can generate tremendous instability."

Glaciologists also know that by itself, free-floating sea ice does not raise the level of the sea, just as an ice cube in a glass of water does not cause an overflow as it melts. But glaciers are different because they rest on land, and if that vast volume of ice slides into the sea at a high rate, this adds mass to the

Glaciers in Retreat

Antarctic researchers are using space-based laser and radar imaging to map and monitor the movement of its ice sheets. Below, ICESat, a NASA satellite, shoots laser pulses to measure ice height. These large-scale studies are complemented by aircraft measurements in regions of greatest change.

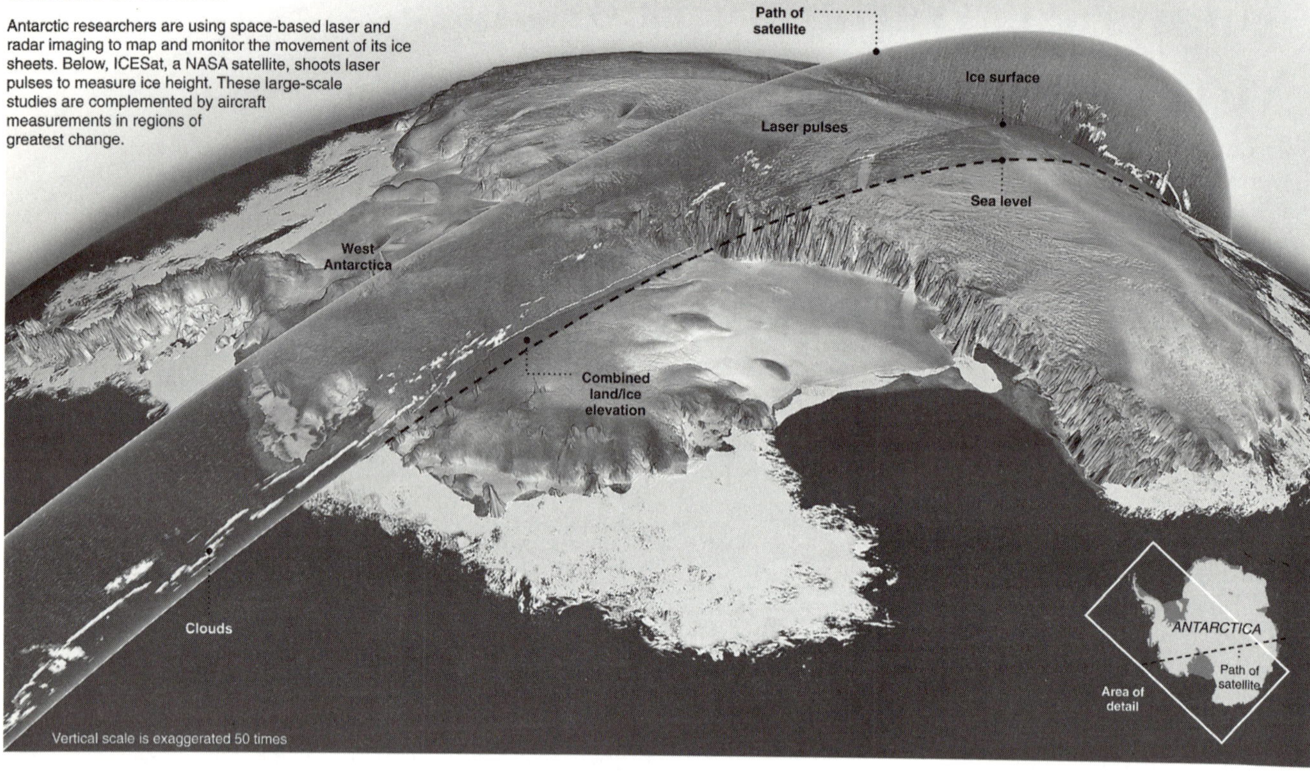

Western Ice Sheets Are Shrinking ...

Recent satellite data suggest that the Pine Island Glacier in West Antarctica is thinning rapidly. From 1992 to 2000, the grounding line - the line where ground-based ice turns to floating ice - retreated up to three miles.

······· 1992 —— 2000

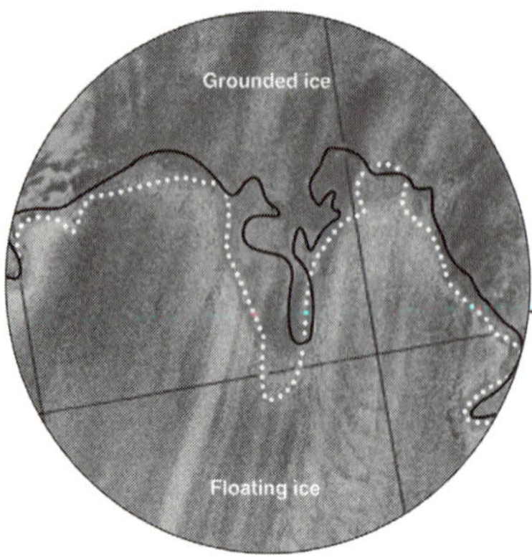

... Partly Because of Low Land Levels ...

While the East Antarctic ice sheet covers mostly land at or above sea level, much of the land under the western ice sheet is below sea level - making the ice susceptible to –accelerated flow into the ocean as its environment changes.

Above sea level   Below sea level

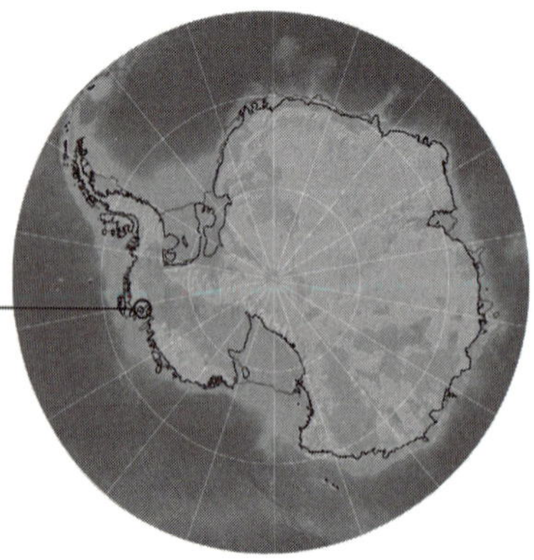

... and Rising Temperatures

Although most of Antarctica has experienced cooling in the last 20 years, satellite data show warming along the coasts, which is believed to be weakening and thinning the ice.

Change in average surface temperature per decade   0   +0.5C   1C   1.5C

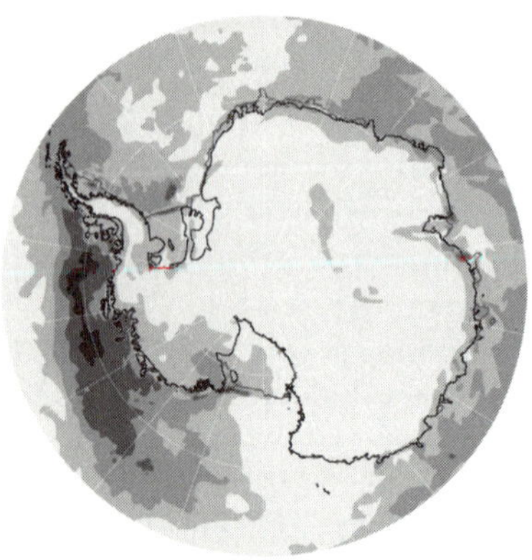

Sources: Waleed Abdalati, Dr. Josefino C. Comiso and the Scientific Visualization Studio, NASA Goddard Space Flight Center; Eric Rignot, NASA Jet Propulsion Laboratory; Canadian Space Agency

David Constantine/The New York Times

ocean, which in turn can raise the global sea level.

Through their flights over this and other areas of Antarctica, NASA and the Chilean center hope to help glaciologists and other scientists interested in climate change understand what is taking place on the continent and why. To do that, they need to compile data not only on ice thicknesses but also the underlying geology of the region, information that is most easily obtained from the air.

The flights are taking place aboard a Chilean Navy Orion P-3 plane that has been specially equipped with sophisticated instruments. The devices include a laser-imaging system that shoots 5,000 pulses of light per second at the ground to map the ice surface, as well as ice-penetrating radar to determine the depth of the ice sheets, a magnetometer and digital cameras.

For most parts of Antarctica, reliable records go back less than 50 years, and data from satellites and overflights like the ones going on here have been collected over only the past decade or so. But that research, plus striking changes that are visible to the naked eye, all point toward the disturbance of climate patterns thought to have been in place for thousands of years.

In 1995, for instance, the Larsen A ice shelf disintegrated, followed in 1998 by the collapse of the nearby Wilkins ice shelf. Over a 35-day period early in 2002, at the end of the Southern Hemisphere summer, the Larsen B ice shelf shattered, losing more than a quarter of its total mass and setting thousands of icebergs adrift in the Weddell Sea.

"The response time scale of ice dynamics is a lot shorter than we used to think it was," said Dr. Robert Bindschadler, a NASA scientist who is director of the West Antarctic Ice Sheet Initiative. "We don't know what the exact cause is, but what we observe going on today is likely to be what is also happening tomorrow."

Thus far, all of the ice shelves that have collapsed are located on the Antarctic peninsula. In reality a collection of islands, mountain ranges and glaciers, the peninsula juts northward toward Argentina and Chile and is "really getting hot, competing with the Yukon for the title of the fastest warming place on the globe," in the words of Dr. Eric Steig, a glaciologist who teaches at the University of Washington.

According to a recent study published in *Geophysical Research Letters*, the discharge rate of three important glaciers still remaining on the peninsula accelerated eightfold just from 2000 to 2003. "Ice is thinning at the rate of tens of meters per year" on the peninsula, with glacier elevations in some places having dropped by as much as 124 feet in six months, the study found.

But the narrow peninsula contains relatively little inland ice. Glaciologists are more concerned that they are now beginning to detect similar signs closer to the South Pole, on the main body of the continent, where ice shelves are much larger—and could contribute far more to sea level changes. Of particular interest is this remote and almost inaccessible region known as "the weak underbelly of West Antarctica," where some individual ice shelves are as large as Texas or Spain and much of the land on which they rest lies under sea level.

"This is probably the most active part of Antarctica," said Dr. Eric Rignot, a glaciologist at the Jet Propulsion Laboratory in Pasadena, Calif., and the principal author of the Geophysical Research Letters paper. "Glaciers are changing rapidly and increasingly discharging into the ocean, which contributes to sea level rise in a more significant way than any other part of Antarctica."

According to another paper, published in the journal *Science* in September, "the catchment regions of Amundsen Sea glaciers contain enough ice to raise sea level by 1.3 meters," or about four feet. While the current sea level rise attributable to glacier thinning here is a relatively modest 0.2 millimeters a year, or about 10 percent of the total global increase, the paper noted that near the coast the process had accelerated and might continue to do so.

As a result, the most recent flights of NASA and the Chilean center have been directed over the Thurston Island and Pine Island zones of West Antarctica, near the point where the Bellinghausen and Amundsen Seas come together. The idea is to use the laser and radar readings being gathered to establish a base line for comparison with future measurements, to be taken every two years or so.

"We're not sure yet how to connect what we see on the peninsula with what we observe going on further south, but both are very clearly dramatic and dynamic events," Dr. Bindschadler said. "On the peninsula, large amounts of melt water are directly connected to disintegration of the ice shelf, but the actual mechanism in West Antarctica, whether melt water, a slippery hill or a firmer bedrock, is not yet clear. Hence the need for more data."

The information being gathered here coincides with the recent publication of a report on accelerating climate change in the Arctic, an area that has been far more scrutinized than Antarctica. That study, commissioned by the United States and seven other nations, found permafrost there to be thawing and glaciers and sea ice to be retreating markedly, raising new concerns about global warming and its impact.

"The Arctic has lots of land at high latitudes, and the presence of land masses helps snow melt off more quickly," said Dr. Steig. "But there's not a lot of land to speak of in the high latitudes of the Southern Hemisphere," making the search for an explanation of what is going on here even more complicated.

The hypotheses scientists offer for the causes of glacier and ice shelf thinning in Antarctica are varied. Rising air, land and ocean temperatures or some combination of them have all been cited.

Some scientists have even proposed that a healing of the seasonal ozone hole over the South Pole and southernmost Chile, a phenomenon expected to take place in the next 50 years or so, could change the circulation of the atmosphere over the frozen continent in ways that could accelerate the thinning of Antarctic ice fields. But even without that prospect, the situation developing in Antarctica is already sobering, glaciologists agree. The data being collected here in West Antarctica and on the peninsula farther north make that obvious, they say, though the degree to which that should be cause for concern around the rest of the planet will become clear only with more research.

"If Antarctica collapses, it will have a major effect on the whole globe," Dr. Rignot cautioned. He warned that "this is not for tomorrow, and Antarctica is such a big place that it's important to look at other areas" around the perimeter of the giant continent, but added, "Nature is playing a little experiment with us, showing us what could happen if the plug were to be removed."

Correction: January 31, 2005, Monday. An article in *Science Times* on Tuesday about the effects of warming in Antarctica misidentified a feature of the landscape there. What is as large as Texas and sits on land that lies below sea level is ice sheets, not ice shelves.

IN REVIEW

1. List the changes that are happening in Antarctica due to increases in average surface temperatures on the continent.
2. Explain how warming in Antarctica might increase the average sea level on Earth.
3. Describe how scientists are measuring changes in the amount of water ice in Antarctica.
4. In what ways are the findings from Antarctica surprising to scientists studying global warming?
5. Does the information in this article alter your opinion in any way about what we should do, if anything, about the issue of global warming? Explain.

Although originating about 150 million kilometers from the Earth, releases of high-energy particles from the Sun, known as coronal mass ejections, can have significant impacts on our home planet. For example, ejections with strong solar storms have damaged satellites, disrupted radio communications, and caused electrical outages. As the world comes to rely on these technologies more and more, several U.S. government agencies have combined resources to develop the National Space Weather Program. Using such satellites as the Solar and Heliospheric Observatory (SOHO), scientists are able to warn industry and government agencies to protect sensitive assets. Because these energetic solar particles move at such tremendous speeds, warnings from SOHO may provide only a few hours of lead time in which to take preventive action.

A Huge Solar Storm, but Little Impact Is Seen

By Anahad O'Connor and Matthew L. Wald
***The New York Times*, October 30, 2003**

One of the largest solar upheavals ever recorded bombarded Earth yesterday with a speeding cloud of charged particles, prompting officials in the United States and Canada to warn airline passengers and crew members that they could be exposed to abnormally high levels of radiation.

Aside from that warning, there were few disruptions from a continuing event that has the potential to interfere with air travel, telecommunications and electricity generation over much of the planet.

The Federal Aviation Administration warned passengers on aircraft over 25,000 feet north of the 35th parallel, which passes through Chattanooga, Tenn., and Albuquerque, that they would accumulate about two millirems of radiation per hour, or two days' worth of normal radiation exposure on the ground.

Operators of electric power systems across the continent, meanwhile, took precautions to prevent the storm from causing blackouts. Only minor effects were reported, said Larry Combs, a space weather forecaster with the National Oceanic and Atmospheric Administration in Boulder, Colo.

"The power companies are trying to avoid operating over capacity," Mr. Combs said.

The storm was not as severe as one in July 2000, experts said. But under an alert system instituted this summer by the F.A.A., the agency's Civil Aeromedical Institute, in Oklahoma City, issued an alert at 3:20 p.m., Central time. It also issued an alert on Tuesday. This was the first time the system had been used, said Wallace Friedberg, the supervisor of the radiobiology section of the institute. The alert did not require any action by any airline, however.

Still, Dr. Robert J. Barish, a medical radiation specialist in New York whose 1996 book "The Invisible Passenger" deals with in-flight radiation, said the exposure to passengers and crew members was at a level that would be subject to government regulation if it occurred on the ground. Instead of the equivalent of a chest X-ray for every three or four hours of flight, he said, "it's two every hour."

The storm, traveling nearly five million miles an hour, hit Earth yesterday at about 1 a.m., Eastern time, and was given the highest rating on the atmospheric administration's geomagnetic storm scale, a G5. Around 4 p.m., Eastern time, scientists saw another giant flare erupt from the same region of sunspots that unleashed the first storm. The Sun has not rotated enough to keep the second storm from heading toward Earth, so there could be a layering effect when it arrives sometime today or early Friday, said Dr. John Kohl, an astrophysicist at the Harvard-Smithsonian Center for Astrophysics.

"We're going to be seeing pretty high levels of radiation over the coming days," Dr. Kohl said.

The initial impact of yesterday's storm, which is expected to last through today, was minimal because its magnetic field was aligned with Earth's, or pointed northward. But for several hours, beginning at about 1 p.m., Eastern time, it became oriented southward, scientists said.

The shift shows that the storm's strength could be growing, said Dr. Devrie Intriligator, director of the space Plasma Laboratory at the Carmel Research Center in Santa Monica, Calif.

Astronauts aboard the International Space Station were urged to take refuge in the craft's living quarters from the blasts of radiation.

And early yesterday, Japan's space agency reported that its Kodama communications satellite malfunctioned and had to be turned off until the storm subsided.

IN REVIEW

1. List the measures that were taken to protect sensitive equipment and the public during the large solar storm in October 2003.
2. How might airline passengers be affected by solar storms when people on the Earth's surface are not affected?
3. Some large coronal mass ejections don't impact the Earth. Explain why not.
4. Astronauts aboard the International Space Station were ordered to take shelter during a large solar storm. What impacts would large solar storms have on future explorers of the Moon and Mars? How could the impacts on these future explorers be lessened?

For more than three decades, the "solar neutrino problem" was one of the great mysteries of the Sun. As discussed in the text, it now appears to have been solved, thanks to data collected by the Sudbury Neutrino Observatory (SNO) and other neutrino detectors. With this success in the bag, scientists are now looking to learn much more from neutrinos—not just about the Sun, but about processes throughout the universe.

Tiny, Plentiful and Really Hard to Catch

By Kenneth Chang
***The New York Times,* April 26, 2005**

An hour north of Duluth, Minn., and a half-mile down, the dim tunnels of the Soudan mine open up to a bright, comfortably warm cavern roughly the size of a gymnasium, 45 feet high, 50 feet wide, 270 feet long.

Well hidden from the lakes, pine forests and small towns of northern Minnesota, the mine churned out almost pure iron ore until it closed in 1962. Today, it is a state park, and it houses a $55 million particle physics experiment that is part of a worldwide effort to unravel the secrets of the neutrino, one of the least known and most common elementary particles.

Because of discoveries over the past decade, the ubiquitous neutrino, once a curiosity in a corner of particle physics, now has the potential to disrupt much of what physicists think they know about the subatomic world. It may hold a key to understanding the creation of hydrogen, helium and other light elements minutes after the Big Bang and to how dying stars explode.

The experiment at Soudan will measure the rate that neutrinos seemingly magically change their types, giving physicists a better idea of the minute mass they carry. An experiment at Fermilab outside Chicago is looking for a particle called a "sterile neutrino" that never interacts with the rest of the universe except through gravity.

Astrophysicists are building neutrino observatories in Antarctica and the Mediterranean, which will provide new views of the cosmos, illuminating the violent happenings at the centers of galaxies, distant bright quasars and elsewhere.

The particle is nothing if not elusive. In 1987, astronomers counted 19 neutrinos from an explosion of a star in the nearby Large Magellanic Cloud, 19 out of the billion trillion trillion trillion trillion neutrinos that flew from the supernova. The observation confirmed the basic understanding that supernovas are set off by the gravitational collapse of stars, but there were not enough data to discern much about the neutrinos.

The much larger detectors in operation today, Super-Kamiokande in Japan, filled with 12.5 million gallons of water, and the Sudbury Neutrino Observatory in Canada, would capture thousands of neutrinos from a similar outburst.

Because neutrinos are so aloof, successful experiments must have either a lot of neutrinos, produced en masse by accelerators or nuclear reactors, or a lot of matter for neutrinos to run into. Given the cost of building huge detectors, scientists are now turning to places where nature will cooperate.

In Antarctica, the IceCube project will consist of 80 strings holding 4,800 detectors in the ice, turning a cubic kilometer of ice into a neutrino telescope. Fourteen European laboratories are collaborating on a project called Antares that will similarly turn a section of the Mediterranean off the French Riviera into a neutrino detector.

The Soudan experiment takes the other approach, using bountiful bursts of neutrinos generated by a particle accelerator. Shoehorned into the back of the underground cavern is a detector of modest size, a mere 6,000 tons, consisting of 486 octagonal steel plates standing upright like a loaf of bread. Each plate, 1 inch thick and 30 feet wide, weighs 12 tons.

On a visit to the cavern last month, William H. Miller, the laboratory manager, pointed at the far rock wall. "Fermilab, that way," he said. This experiment is intended to catch just a few of the neutrinos created at Fermilab, 450 miles away, which gush out of the rock wall, through the cavern, through the steel plates and then through another several miles of rock before emerging out of the earth and continuing into outer space, having no effect on Dr. Miller or the reporter interviewing him.

"You need a light-year of lead to reliably stop" a neutrino, said Dr. Alec T. Habig, a professor of astrophysics at the University of Minnesota at Duluth and the operations manager for the neutrino detector.

Only occasionally, a neutrino runs into a proton or neutron among the many atoms in the steel plates, and the wreckage of that collision is recorded as tiny bursts of light careering through the detector.

When the experiment begins running at full speed later this year, Fermilab will send trillions of neutrinos every couple of seconds flying toward Soudan. The beam will spread out to half a mile wide by the time it reaches Minnesota, so most of the neutrinos will miss the cavern entirely. But even among those that strike the bull's-eye, only one every few hours will actually hit something in the detector and be detected.

So far, in the testing phase in the past two months, the Soudan detector has seen just three, maybe four, neutrinos from Fermilab. But then, Wolfgang Pauli thought physicists would never see any. Pauli, a pioneer of quantum theory, contrived the notion of neutrinos in 1930 to explain the disappearance of energy when unstable atoms fell apart. Pauli said the missing energy was being carried away by an unseen particle.

"I've done a terrible thing," Pauli wrote. "I have postulated a particle that cannot be detected." Pauli even wagered a case of Champagne that his particle would not be detected. In 1956, Pauli sent a case of Champagne to Clyde L. Cowan Jr. and Frederick Reines, two American physicists who proved him wrong using the flood of neutrinos produced in a nuclear reactor.

Physicists later discovered that neutrinos come in three types, whimsically called flavors. The flavor seen first was the electron neutrino, which interacts only with electrons. Heavier electron-like particles known as muons and tau particles are accompanied by their own flavors of neutrinos.

In 1998, an experiment at Super-Kamiokande showed that neutrinos change flavors as they travel along. For that to occur, the laws of physics dictate that the neutrinos, which had been thought be massless, must actually carry along a smidgen of weight, less than a millionth as much as an electron, the next lightest particle. Each flavor also has a slightly different mass.

In the Fermilab-to-Soudan experiment, the neutrinos are generated from a beam of protons, which are directed down a newly built $125 million tunnel, focused to a very narrow width with powerful electric fields and then slammed into a piece of graphite. That produces short-lived particles called pions, which in turn generate muon neutrinos as they decay.

The beam passes through a smaller version of the Soudan detector, allowing the physicists to verify the number of neutrinos. The tunnel, sloped downward three degrees, ends just beyond the detector. The neutrinos keep going, into the earth, to emerge in the Soudan cavern one four-hundredth of a second later.

By using neutrinos created in an accelerator, physicists will be able to vary the energy of the neutrinos and see how that changes the number detected at Soudan.

"This wiggle has not really been seen," said Dr. Boris Kayser, a physicist at Fermilab. "It is one of the central expectations of our picture of neutrino oscillations."

The data from Soudan is expected to refine the Super-Kamiokande results, not overturn the prevailing wisdom. Another experiment at Fermilab may do just that.

A decade ago scientists at Los Alamos National Laboratory in New Mexico looked at neutrinos traveling a short distance from a nuclear reactor and saw indications of a large oscillation, suggesting a relatively large mass gap between two of the neutrino flavors. The gap was large enough that it could not fit into any theory consisting of just three neutrinos. But other experiments showed that only three flavors of neutrinos that interact with ordinary matter exist.

That has led to speculations of a new class of particles called sterile neutrinos. These particles would exert a force on other matter through gravity but would otherwise be completely inert.

"If it's really due to oscillations, then it implies physics way beyond the Standard Model," said Dr. William C. Louis of Los Alamos, who worked on the experiment.

Dr. Kayser said most theorists "are skeptical, because it doesn't fit," yet no one can point to obvious flaws in the Los Alamos work, either. The new experiment at Fermilab, called MiniBooNE, is looking for the same effect but with a different setup, firing neutrinos into a spherical tank containing 250,000 gallons of baby oil.

(The name BooNE is an awkward contraction of Booster Neutrino Experiment. Booster refers to Fermilab's booster ring that accelerates protons, and the project's leaders added the prefix "mini" because they imagined a second, larger stage with a second detector if the current "mini" run confirms the Los Alamos findings.)

Dr. Louis, who is also one of the spokesmen for MiniBooNE, said that initial answers could be out by fall and insisted that he was not betting either way. "We're just concentrating on getting the correct result," he said, "and we'll worry about the consequences later."

The consequences may include the understanding of atom production in the aftermath of the Big Bang and in supernovas. Because neutrinos are essential to the nuclear reactions that change protons to neutrons and vice versa, they influence which elements form in what relative proportions.

"That would have profound implications for our models of the early universe and for supernovas," said Dr. George M. Fuller, a professor of physics at the University of California, San Diego. "It could change everything."

The problem is that current models that include three flavors of neutrinos do a good job of explaining the amount of hydrogen and helium in the universe. The existence of sterile neutrinos would send astrophysicists scurrying to come up with new calculations to produce the same answers.

On the other hand, current models of supernovas have trouble producing enough neutrons to form the heavier elements like uranium, and Dr. Fuller said sterile neutrinos could shift the reactions toward producing more neutrons.

Future experiments should aim at understanding other aspects of neutrinos, said Dr. Kayser, who was co-chairman of a committee that just released recommendations for future neutrino study.

For one, the neutrino oscillation findings say only that a difference in mass between the different flavors exists, but not the exact mass of any of them. The presumption is that because an electron is lighter than a muon and a muon is lighter than a tau that the same pattern should be true of the three neutrino flavors, with the electron neutrino the lightest and the tau neutrino the heaviest. But that does not have to be the case.

"It could be the other way around," Dr. Kayser said.

Physicists are also trying to learn whether an antineutrino is actually a neutrino. (Other antiparticles have opposite electrical charge. Because neutrinos are electrically neutral, nothing would prevent a neutrino from being its own antiparticle.)

Another open question is whether neutrinos play a role in the imbalance of matter and antimatter. If the early universe had contained equal amounts of the both, everything would have been annihilated, leaving nothing behind to form stars and galaxies.

Among quarks, which form protons and neutrons, physicists have observed a subtle matter-antimatter imbalance, called CP violation, in the behavior of particles known as mesons. "That CP violation is completely inadequate to explain the universe that we see," Dr. Kayser said.

So physicists suspect that there must be CP violation elsewhere and that the oddity of neutrinos suggest they could be a source. That, in turn, leads to speculation of yet more new types of neutrinos—very heavy ones that existed only in the very early universe—and the decay of those heavy neutrinos created the preponderance of matter.

Then come even wilder ideas—that neutrinos play a role in the mysterious dark energy that is pushing the universe apart or that neutrinos could be used for interstellar communication.

"Most of these ideas are of course probably wrong," said Dr. Louis of Los Alamos. "But if even one of them is right, it would be a tremendous breakthrough."

IN REVIEW

1. What are neutrinos? Why are they so difficult to catch? Why are neutrino detectors built deep underground?
2. Aside from solar emissions, what other cosmic events have produced neutrinos that we have detected in the past?
3. Briefly describe the Soudan neutrino experiment. What do scientists hope to learn from it? How will it complement other neutrino experiments, such as the Sudbury Neutrino Observatory?
4. How might a better understanding of neutrinos and observations of cosmic neutrinos lead to a better understanding of the early universe?
5. Until recently, we were able to study the distant cosmos only by observing light. How are neutrino observatories opening up a new window on the universe?

As discussed in the text, stars come in a wide range of masses and an even wider range of luminosities. But theoretical modeling suggests an upper limit on the masses of stars, and the superstar described in this article appears to have a mass near this upper boundary.

Astronomers Call LBV 1806-20 the Superstar That Outshines All

By Kenneth Chang
***The New York Times,* January 6, 2004**

Correction Appended

Astronomers have spotted what they believe is the biggest, brightest star they have ever seen—and indeed bigger than can be explained by the usual theories of how stars form.

The star, LBV 1806-20, is 5 million to 40 million times as bright as the Sun, at least 150 times as massive and at least 200 times as wide, astronomers reported on Monday at a meeting of the American Astronomical Society here.

Next to LBV 1806-20, the Sun would appear just as minuscule as Mercury does next to the Sun.

Dr. Steven Eikenberry, a professor of astronomy at the University of Florida, who led the research, said it was surprising that even after decades of observing the Milky Way, "we're still finding these big monsters out there."

The findings have been submitted to The Astrophysical Journal.

The brightest star known until now is one called the Pistol Star, five million to six million times as bright as the Sun. The brightness of a star, or its power output, is proportional to the cube of its mass; a star 10 times the mass of the Sun would be 1,000 times as bright, and a star 100 times the mass would be one million times as bright.

LBV 1806-20, despite its brightness, is not easily seen. Forty-five thousand light-years away on the other side of the Milky Way galaxy, it is blocked from view by dust clouds. But about 10 percent of its infrared light does make it to Earth.

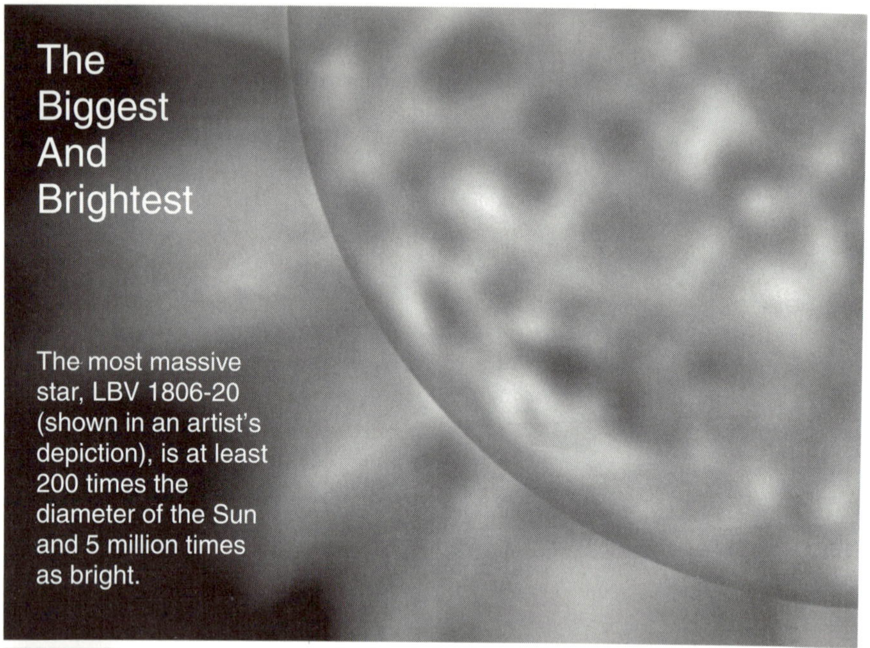

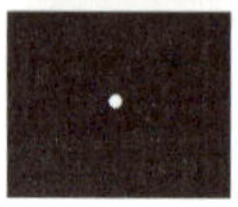
The size of the Sun, left, at the same scale as LBV 1806-20, above.

Source: University of Florida

The New York Times

When LBV 1806-20 was discovered in the 1990's, astronomers categorized it as a bright, short-lived blue star and estimated that it was at least a million times the mass of the Sun. But with new observations from the Palomar Observatory in California and the Cerro Tololo Inter-American Observatory in Chile, astronomers produced better images and new estimates of mass and brightness. Because the star is so bright, it will burn out quickly, within a few million years, compared with the 10 billion-year lifespan estimated for the Sun.

Dr. Eikenberry said the high-resolution images ruled out the possibility that fooled astronomers previously into erroneous claims: that the supposed supermassive star was actually a cluster of smaller ones. Still, he said it might be a binary or triple-star system.

If it is a single star, it probably did not form the usual way: a cloud of hydrogen gas collapsing under gravity, heating up and turning on when the hydrogen fuses into helium. But calculations indicate that stars should be at most 120 times as massive as the Sun, because heat and radiation from the collapsing cloud would blow away any more gas.

A clue to LBV 1806-20's history can be found in its neighbors, Dr. Eikenberry said. Nearby are other stars, which, while not quite reaching LBV 1806-20's girth, are still very large; there are also a proto-star and a neutron star, the burned-out core of a star that exploded in a supernova.

"Nothing like this has ever been seen before," Dr. Eikenberry said.

Perhaps when the star exploded, the shock wave caused nearby gas clouds to collapse and form the large stars, including LBV 1806-20, he said. When these stars exhaust their hydrogen fuel in a few million years, they, too, will explode in a succession of supernovas, he went on, "like a fireworks display."

Correction: January 7, 2004, Wednesday. An article yesterday about the biggest, brightest star ever observed misstated an earlier assessment of it. When the star was discovered in the 1990's, astronomers estimated that it was at least a million times the brightness, not the mass, of the Sun.

IN REVIEW

1. Describe why the star, LBV 1806-20, is unusual. What are its basic properties?
2. How does LBV 1806-20 compare to the Pistol Star? If LBV 1806-20 is indeed bigger than the Pistol Star, why did it take longer to discover?
3. How does the luminosity of a star relate to the total time it will live? Explain how a star's mass affects its total lifetime, and specifically, how its mass affects nuclear fusion in its core.
4. Why are astronomers uncertain as to whether LBV 1806-20 is a single star or a binary or triple star system? What will be required for astronomers to determine if it is, in fact, a single star?
5. Explain how the nearby stars give clues to the history of LBV 1806-20.

Using human life as an example, a Sun-like star is born, goes through a turbulent adolescent phase as a proto-star, settles down as an adult through a very long and stable middle age as a main sequence star, eventually ages as a bloated red giant, and dies as a planetary nebula surrounding a white dwarf. How long stars live and the manner of their death ultimately depends on the mass of the dust cloud from which they are born: Their fate is sealed from before birth. Stars much more massive than the Sun burn bright and die young, in a spectacular supernova. Less massive stars steadily shine at a dimmer level, enduring at a more leisurely pace, and end their lives more gently, as white dwarfs in a planetary nebula phase. However, recent observations indicate that stars, unlike humans, may have a second chance for extended life when two of them collide, merging into one star. The merger resuscitates mature stars, renewing the hydrogen fusion within the core. Such regeneration would explain a mystery that has plagued scientists for years.

Two Stars Collide; A New Star Is Born

By Kenneth Chang
***The New York Times,* June 13, 2000**

With most of the universe being empty space, astronomers once thought the chance of stars colliding was roughly once in an eternity.

But in recent years, astronomers have realized that collisions are not uncommon within dense clusters of stars.

"We believe there are several hundred every hour somewhere in the universe," said Dr. Michael Shara, curator of the astrophysics department at the American Museum of Natural History, in a symposium at the museum two weeks ago.

At the meeting, the first devoted to interstellar collisions, astronomers said they had detected what appeared to be the aftermath of a three-star collision. Scientists are also hoping to find evidence of collisions involving ultradense neutron stars and black holes by detecting telltale gravitational ripples.

The first clue of interstellar collisions came in the 1950's when astronomers looked at clusters of stars they knew to be nearly as old as the universe and saw what appeared to be large, young blue stars. Blue stars contain more hydrogen than smaller stars, but burn hotter and burn out more quickly.

Within the star groups, known as globular clusters, the clouds of gas and dust that give birth to stars had been exhausted billions of years ago, which raised the question: where did these young blue stars come from?

At first, some astronomers speculated that these "blue straggler" stars had somehow conserved their fuel, but with the help of the Hubble Space Telescope, astronomers have concluded that a blue straggler star is actually two small, old stars that combined to form a rejuvenated blue star.

Collisions are more frequent in globular clusters, because there are many more stars. In the outer region of the galaxy where our solar system lies, the density of stars is about one per 10 cubic light-years. In a globular cluster, there may be several hundred thousand stars in that same space.

Dr. Shara's estimate of several hundred collisions per hour may seem frequent, but with 100 billion galaxies in the observable universe and each galaxy containing, on average, 30 globular clusters, most of the collisions occur far away and out of view.

According to his calculations, over the 10 billion-year lifetime of the Milky Way, Earth's home galaxy, there have been at least one million collisions within globular clusters—or about one every 10,000 years.

While astronomers can spot the results of the collisions, the blue straggler stars, they almost certainly will never see the collisions.

This means that they cannot directly determine whether a particular blue straggler star originated in a binary system whose components coalesced or is the wreckage of two unrelated stars that happened to collide.

In one case, however, the answer appears to be the latter.

At the symposium, Dr. Rex Saffer, an assistant professor of astronomy and astrophysics at Villanova University, presented observations of several blue straggler stars within the globular cluster NGC 6397, located in the Milky Way about 7,000 light-years from Earth.

Based on the age of the cluster, its surviving stars should have masses of no more than 80 percent of the mass of the Sun; more massive stars should have burnt out.

However, four of the blue stragglers in the cluster were about 1.6 times the mass of the Sun, so they presumably represent the mergers of two stars.

One blue straggler was a whopping 2.4 times heavier—"so massive that it can't have formed by the merger of two stars," Dr. Saffer said. That straggler, he said, was probably born from a three-star collision, when an outsider star flew into a binary system.

"The stars go into a chaotic dance around each other," he said, and eventually bump into one another.

Dr. Shara described the observations as "almost certainly irrefutable proof" that this particular blue straggler formed in a collision rather than the simple coalescing of a binary system. (The latter phenomenon almost certainly occurs; astronomers have also seen binary stars so close that they are touching each other.)

If telescopes cannot actually watch collisions, computers offer some insight into what an interstellar pileup might look like.

Describing a simulation in which one star crashes into another about twice its size at a speed of about half a million miles per hour, "the small one actually plows through the larger one and tears a furrow," said Dr. Joshua E. Barnes, an associate professor of astronomy at the University of Hawaii.

The crash sends a splash of gas into space while the smaller, denser star burrows to the center of the larger one.

"There's a huge wave that gets set up in the larger one, and it's going to splash around for while," Dr. Barnes said. "The bigger one is wobbling around."

What happens after that is not clear, because the simulation only covers the first 12 hours.

"After that initial violence dies down, you've got an object that in many respects is not a proper star," Dr. Barnes said. "It's all swelled up, and internally, its temperature structure is not what you expect for a star. Getting through this intermediate stage where this new formed blob is sorting itself out and trying to turn back into a star, that's a long process" lasting from 100,000 years to 10 million years.

Even near misses can leave a mark. The most common stars within globular clusters are the low-mass stars and white dwarfs, the cores of burned-out stars.

Dr. Adrienne Cool, an associate professor of physics and astronomy at San Francisco State University, and her collaborators surveyed stars near the core of NGC 6397, looking for instances where a near collision between a star and a white dwarf ended up with the two in orbit around each other. The result is a "cataclysmic binary," where the white dwarf flares intermittently as gas sucked from the other star falls onto it.

"We found a handful of them in this cluster," Dr. Cool said.

The researchers also found the unexpected: mini-white dwarfs with half the mass of usual white dwarfs. Dr. Cool hypothesizes that these are remnants of red giant stars that had close encounters with other stars.

As a star exhausts its hydrogen, astronomers say, it swells into a red giant, expanding, as would be the case with the Sun, past the orbit of Earth, before collapsing into a white dwarf.

In the middle of a red giant is the seed of a white dwarf. A near miss with another star could rip away the outer layers. "If you can somehow strip off that red giant's envelope early," Dr. Cool said, "then you can lay bare this unusual core, this very low-mass white dwarf."

Other participants at the symposium are exploring the more cataclysmic collisions of neutron stars and black holes.

Such collisions are believed to create gravitational disturbances that may be detectable in an experiment called the Laser Interferometer Gravitational Wave Observatory, or LIGO. Initial observations should begin next spring with full operations in 2002.

According to Einstein's general theory of relativity, as gravitational waves pass, the distance between any two points contracts and expands. LIGO's two detectors—one in Hanford, Wash., the other in Livingston, La.—will be able to spot a change as slight as half the width of an atom along the 2.5-mile-long paths traveled by the detectors' laser beams.

That should be sensitive enough to pick out ripples emanating from the crash of two neutron stars up to 100 million light-years away or a collision of black holes up to 500 million light-years away.

Theorists have been trying to determine whether these collisions occur often enough and close enough for LIGO to spot.

If recent calculations are right, probably not. "We don't have any comfortable predictions to expect a detection with LIGO I," said Dr. Vassiliki Kalogera, a post-doctoral fellow at the Harvard-Smithsonian Center for Astrophysics.

At best, Dr. Kalogera said, LIGO will detect one collision of black holes in a year.

The predictions are more optimistic for an upgraded LIGO, scheduled for 2006, that will be able to detect waves from collisions 10 times farther away.

Then, Dr. Kalogera said, a "conservative lower limit" would be the detection of 1 or 2 neutron star collisions and 10 black hole collisions per year.

"We could have a couple of hundred events per year if our most optimistic predictions are correct," Dr. Kalogera said.

IN REVIEW

1. Explain the mystery that is potentially solved by stellar collisions.
2. With the distance between stars being so vast, how could several hundred stellar collisions occur every hour? Why would globular clusters be a more common place for stellar collisions?
3. What specific evidence do scientists have that a new star may have formed from two stars colliding?
4. How does a new star form when two stars collide? What happens in the interior of the newly formed star that rejuvenates it?
5. The article discusses other types of collisions. Explain how these collisions are different from a collision of two stars to form a new star.

This *New York Times* article discusses Type 1a supernovas, which the textbook describes as "white dwarf supernovas." As discussed in the text, a white dwarf supernova is thought to occur in a binary star system in which one star is a white dwarf and the other star transfers mass to it. Once the mass of the white dwarf exceeds the limit of 1.4 times solar mass (which the article refers to as the "Chandrasekhar mass"), it explodes as a supernova.

Life-or-Death Question: How Supernovas Happen

By Dennis Overbye
***The New York Times,* November 9, 2004**

Once a second or so, somewhere in the universe, a star blows itself to smithereens, blossoming momentarily to a brilliance greater than a billion suns.

Nobody understands how these events, among the most violent in nature, actually happen. But, until recently, that didn't much matter unless you were a practitioner of the arcane and messy branch of science known as nuclear astrophysics.

Lately, however, supernovas have become signal events in the life of the cosmos, as told by modern science.

Using a particular species of supernova, Type 1a, as cosmic distance markers, astronomers have concluded that a mysterious "dark energy" is wrenching space apart, a discovery that has thrown physics and cosmology into an uproar.

As a result, the fate of the universe—or at least our knowledge of it—is at stake, and understanding supernovas has become essential.

Astronomers are busy on many fronts trying to figure out the details of these explosions—scanning the skies to harvest more of them in the act, peering at the remains of ancient supernovas to seek a clue to their demise, harnessing networks of supercomputers to calculate moment by moment reactions in the heart of hell.

This has resulted recently in a kind of two-steps-forward, one-step-back progress, encouraging astronomers that they are on the right track, generally, with their theories, but at the same time underscoring complexities and baffling puzzles when it comes to pinning down the details of what happens in the explosions.

Last month members of an international team of astronomers led by Dr. Pilar Ruiz-Lapuente of the University of Barcelona announced that they had found a star speeding away from the site of a supernova blast seen in 1572 by the astronomer Tycho Brahe. This supernova, which appeared as a "new star" in the constellation Cassiopeia, was one of the earliest studied by astronomers, and helped shatter the Aristotelian notion that the heavens above the Moon were immutable.

The newly discovered star, presumably the companion of the star that exploded, supports a long-held notion that such explosions happen in double star systems when one star accumulating matter from the other reaches a critical mass and goes off like a bomb.

Meanwhile, members of a group of astrophysicists using a network of powerful supercomputers to simulate supernova explosions say they have succeeded for the first time in showing how such a star could blow up.

Over the course of 300 hours of calculation at the University of Chicago's Center for Astrophysical Thermonuclear Flashes, otherwise known as the Flash center, they watched bubbles of thermonuclear fury rise from the depths of the star like a deadly jellyfish and then sweep around the surface and collide in an apocalyptic detonation that Dr. Donald Lamb, a Chicago astrophysicist, called "totally bizarre and novel."

If true, the Chicago results could help explain not only how stars explode, but why the explosions are almost but not exactly alike, allowing astronomers to better calibrate their measurements of dark energy.

Many supernova experts said, however, that such computer simulations were more of a good start than a final answer. Dr. J. Craig Wheeler of the University of Texas called the Flash center work "a courageous calculation," but added that many details needed to be filled in. "I don't think this is the end of the story," he said. The story of Type 1a supernovas, experts have long agreed, begins with a dense cinder known as a white dwarf, composed of carbon and oxygen, which is how moderate-size stars like the Sun, having exhausted their thermonuclear fuels of hydrogen and helium, end their lives.

If it happens to be part of a double star system, the white dwarf can accumulate matter from its companion until it approaches a limit, known as the Chandrasekhar mass—about 1.4 times the mass of the Sun.

At that point, so the story goes, the pressure and density in the previously dead star will be great enough to reignite the star and thermonuclear reactions will ripple upward, transmuting the carbon and oxygen into heavier and heavier elements, ripping the

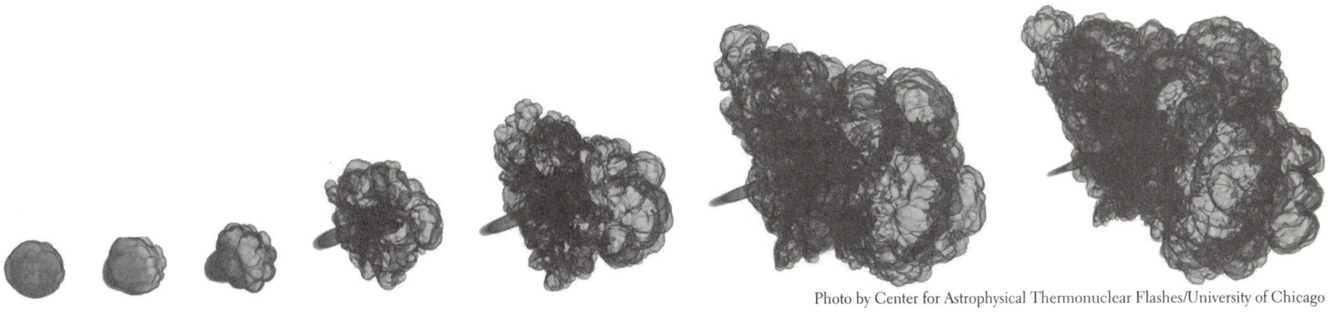

Photo by Center for Astrophysical Thermonuclear Flashes/University of Chicago

A three-billion-degree bubble of thermonuclear hell mushrooms upward through a star in the early milliseconds of a supernova explosion. Sweeping around the star's surface, the bubble could collide with itself, setting off a fatal detonation.

white dwarf apart while its companion goes flying off.

Until recently, however, there was little evidence of this. Two white dwarfs could collide, for example, and blow up. In that case there would be no survivor.

Tycho Brahe's supernova has now offered new evidence for the former model, of the white dwarf bomb.

That supernova is one of the few of Type 1a's that have occurred in our own galaxy, and so astronomers have long sought to find its companion. That star, astronomers reasoned, should be zinging along relative to its neighbors, as a result of having been released, like a stone from a slingshot, from its orbit around the suddenly deceased white dwarf.

The site of the supernova explosion is marked today by a small scruff of X-rays and radio waves in the sky.

Near the center of this patch the team found a sunlike star moving three times as fast as it neighbors.

The star has the right characteristics to have been the one donating material to the white dwarf that exploded, but the identification is not ironclad, a team member, Dr. Alex Filippenko of the University of California at Berkeley, said, explaining in an e-mail message that "it is 'possible' that the star just happened to be zooming through that region and is unrelated to the supernova."

One far out possibility, the astronomers say, is that more observations will reveal the ashes of the supernova polluting the outer layers of the star. But that is probably too much to wish for, said Dr. Stan Woosley of the University of California at Santa Cruz, pointing out that the explosion might have blown the outer layer of the star, ashes and all, off into space.

"This star sat next to, and for a while inside the most powerful thermonuclear explosion in the universe, 2.5 million, trillion, trillion megatons," Dr. Woosley said.

But the details of that explosion, which happens invisibly in a second or so, are still a mystery.

The light show seen by astronomers comes from radiation released by radioactive nickel, which decays to cobalt, and then to iron over the days and months after the cataclysm, releasing gamma rays that strike the ashes of the shattered star and make them glow briefly brighter than a galaxy.

Because all Type 1a supernovas start from the same point, astronomers have tried to use them as cosmic geodetic markers, standard candles whose distances can be inferred from how bright they appear.

But the supernovas are not standard enough. They vary in their luminosities by about 40 percent, which is similar enough to prove that the expansion of the universe is speeding up and that dark energy exists, astronomers say, but not good enough to pin down crucial details about the strength of this strange force and how it may be changing over cosmic time, and thus whether the universe will ultimately rip apart or come together in a "big crunch."

In order to reduce the uncertainties in their measurements astronomers need to know how or whether to correct their observations for differences in things like the age and chemical composition of the parent white dwarfs.

The problem is that there are two ways for the star to burn: like a flame, which is called deflagration, and as an explosion, a detonation, in which the burning propagates as a shock wave.

And neither type of burning, by itself, can easily explain what astronomers have seen in supernova explosions.

The slow burn, deflagration, results in more of a fizzle than an explosion, they say. It does not produce enough nickel to generate the light seen by astronomers and leaves much of the star unburned. Moreover, the parts that are burned are all jumbled up, while supernovas in the sky appear to be nicely layered, with the densest elements, like iron and nickel in the center, and light ones like silicon, sulfur and magnesium on the outside.

If the supernova consists simply of a detonation, on the other hand, the star would all turn to nickel, and that would result in too much light.

As a result, in the last 10 years many theorists have adopted a "Goldilocks" model of the explosion, in which the star burns in the flame mode for a while, slowly expanding, and then detonates when the density of the star has fallen to the value to make the right amount of nickel.

"The porridge has to be just the right temperature," said Dr. Wheeler, who described recent three-dimensional simulations by Dr. Vadim Gamezo and Dr. Elaine Oran, both of the Naval Research Laboratory in Washington, and Dr. Alexei R. Khokhlov of the University of Chicago, as "state of the art."

None of these "delayed detonation" models explain why or when the star would detonate.

The scientists had to put that into the calculation by hand. Finding a natural trigger for the detonation is the "silver chalice" of our profession, Dr. Wheeler said, adding that automobile companies spend millions on the problem of ignition in car cylinders.

This is where the Flash center calculations come in. "It turns out that you need walls to have an explosion," explained Dr. Lamb. But a star has no walls. So how does it explode?

The Flash group, led by Dr. Tomasz Plewa of Chicago and the Nicolaus Copernicus Astronomical Center in Warsaw, was investigating what would happen if the white dwarf began burning in a flamelike manner not exactly at its center—an unlikely event in the case of a real star subject to turbulence—but a bit off-center. In addition to Dr. Lamb, the group included Dr. Alan C. Calder of Chicago.

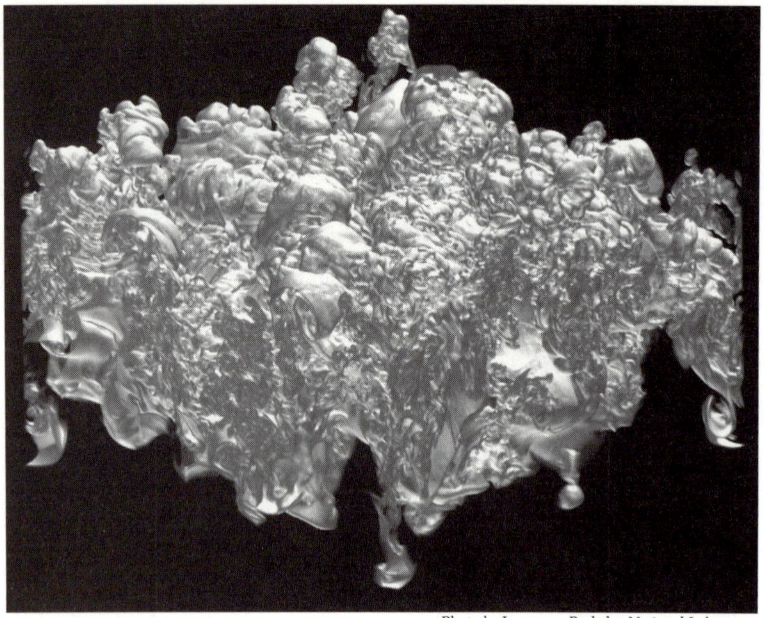

Photo by Lawrence Berkeley National Laboratory

Gravity and buoyancy churn and warp the flame front in a star undergoing a supernova explosion. The front marks the boundary, as thick as a sheet of paper, where oxygen and carbon are being fused to heavier elements. Forty days and 40 nights on a supercomputer were required to produce this image representing a patch about half a yard across.

The result was a bubble of flame rising from the depths and then sweeping around the star to become its own wall, crashing into itself at a temperature of three billion degrees and crushing densities, enough, the Chicago physicists say, to trigger detonation.

"We watched with eyes agog and jaws dropped as the thing unfolded," Dr. Lamb said.

But whether nature really works this way or not, Dr. Lamb and others agree, is yet to be determined, and it is far from a complete theory.

For one thing, the group has not yet been able to make three-dimensional calculations of the actual detonation. Such calculations could be compared to observations.

As Dr. Woosley, said in an e-mail message, "just how a Type 1a supernova explodes is one of the most complicated things in the big wide world."

His group uses supercomputers to study small patches of the turbulent flame front—only a yard or two across—at high resolution.

Dr. David Arnett, a supernova expert at the University of Arizona, said that such simulations were a way to test ideas and that watching them was a prod to theorists' intuition.

"Massive computing does not provide the answers so much as it provides an extension of our imagination," he wrote in an e-mail message. "For some years there has been talk of computing as being the third 'leg' of science: theory, experiment, computer simulations. I think the Flash work is a concrete example of this at work, and actually working."

Meanwhile, real supernovas threaten to confound the theorists. The carbon at the center of the star might "smolder" before it burns, putting it on a path to wind up as something other than nickel at the end, according to recent observations of two supernovas by Dr. Wheeler and his group using the Very Large Telescope at Le Serena, Chile. The evidence is sketchy, but that would mean that most of the models, including the Flash center's rising bubble, are wrong, he said.

But we shouldn't be discouraged. "We've come along way," Dr. Wheeler said. Referring to the ignition problem, he said, "We had to come a long way before we knew this was an issue."

IN REVIEW

1. A white dwarf supernova is one of two basic types of supernova; the textbook refers to the other type as a massive star supernova. How do the two types of supernova differ? Why do astronomers expect white dwarf supernovas all to be similar in luminosity, and why don't they expect the same for massive star supernovas?
2. How are white dwarf supernovas important to establishing the cosmic distance scale? List the main uncertainties associated with measuring the distances from Earth to these supernovas.
3. Describe the tools and observations scientists are using to unravel the mysteries associated with white dwarf supernovas.
4. How has the discovery of a star racing away from the site of Tycho's supernova given support to the idea that these supernovas represent the detonations of white dwarfs? Explain clearly.
5. How is a good understanding of these supernovas relevant to the question of the fate of the universe?

Our Sun's total lifetime is estimated to be about 10 billion years. For most of its life, the Sun will be classified as a main sequence star, where hydrogen is steadily fused to helium in the depths of its core. However, at the end of the Sun's life, it will swell to become a red giant before finally expelling its outer layers as a planetary nebula.

3 Giant Stars May Point the Way to Our Sun's Destiny

By Kenneth Chang
***The New York Times,* January 11, 2005**

KW Sagitarii is not the brightest star in the Milky Way, or the heaviest or densest. Neither are two other stars, KY Cygni and V354 Cephei.

But these three red supergiant stars have caught the attention of astronomers because they appear to be the biggest stars known, and possibly the biggest possible for a normal star. (Another supergiant, VV Cephei, may be even larger, but that one has been distorted by the gravity of a companion star.)

Each of the three stars is more than a billion miles wide, filling as much space as 3.4 billion Suns. Placed at the center of the solar system, one of these stars would extend to halfway between Jupiter and Saturn. The stars are more than twice as wide as Betelgeuse in the constellation Orion, a red supergiant familiar to amateur astronomers.

"We actually know, both observationally and theoretically, how big a star can get," said Dr. Philip Massey of the Lowell Observatory in Flagstaff, Ariz., who led an international team of astronomers that conducted the research. "These are right at the limit of what solar evolutionary theory says we should get by the way of red supergiants."

Emily Levesque, a Massachusetts Institute of Technology undergraduate on the research team, presented the findings on Monday at a meeting of the American Astronomical Society here. A scientific paper describing the results has been submitted to The Astrophysical Journal.

As stars near the end of their lives, they exhaust the hydrogen fuel at their cores. The fusion process that creates light and energy then spreads outward from the core, and the increased pressure pushes the outer layers. When the Sun reaches this stage four billion to five billion years from now, it will probably swell beyond Venus's orbit, becoming a red giant.

KW Sagitarii, KY Cygni and V354 Cephei are much larger, about 25 times as massive as the Sun, and thus swelled far more.

Astronomers do not have celestial rulers to measure the stars, which appear only as points of light to even the most powerful telescopes. Rather, the colors of light emitted by molecules in the outer layers tell the surface temperatures, and models of how a star works produce estimates of the diameters. But observations of red supergiants conflicted with the models.

In the new research, Dr. Massey and his collaborators took new measurements of 74 red supergiants and found that the coolest, and thus largest, had surface temperatures of about 5,750 degrees Fahrenheit, about 10 percent warmer than earlier measurements. "I think everything is now in agreement," he said.

While KW Sagitarii, KY Cygni and V354 Cephei are most likely the largest possible stars in the Milky Way, Dr. Massey said, they may not be the largest in the universe. Stellar models, he said, predict that even larger stars could evolve in galaxies with lower concentrations of elements heavier than helium, like the nearby Magellanic Clouds.

IN REVIEW

1. The Sun will eventually be a red giant, but more massive stars can become red supergiant stars. List the main differences between a red giant star and a red supergiant.
2. How do the scientific theories and astronomical observations about red supergiants agree?
3. Explain why the largest red supergiant stars have the coolest surface temperatures. Explain how this cool temperature affects the color of the star.

When supernova explosions occur, the resulting release of energy creates a flash at its peak luminosity as bright as 10 billion stars combined. But a supernova is not the "mother of all explosions." That title is reserved for gamma ray bursts, whose brightness may, for just an instant, exceed that of a million galaxies. The most recent of NASA's gamma-ray observatories, the Swift satellite, is designed to quickly detect a gamma ray burst and position itself to look squarely down its throat. Launched in November 2004, the Swift satellite has already made significant detections of gamma ray bursts, leaving high-energy astrophysicists salivating.

Dying Star Flares Up, Briefly Outshining Rest of Galaxy

By Kenneth Chang
***The New York Times,* February 20, 2005**

For a fraction of a second in December, a dying remnant of an exploded star let out of a burst of light that outshone the Milky Way's other half-trillion stars combined, astronomers announced Friday.

Even on Earth, half a galaxy away, the starburst was one of the brightest objects ever observed in the sky, after the Sun and perhaps a few comets. The magnitude of the event caught most astronomers by surprise.

"Whoppingly bright," said Dr. Bryan M. Gaensler, an astronomer at the Harvard-Smithsonian Center for Astrophysics in Cambridge, Mass. "It gave off more energy in 0.2 seconds than the Sun does in 100,000 to 200,000 years."

No one on Earth directly saw the flare because most of the light was gamma rays, photons that are more energetic than X-rays and are blocked by the atmosphere. But the Dec. 27 pulse registered on instruments aboard 15 spacecraft, including NASA's new Swift satellite, which was designed to record cosmic gamma rays and had been turned on just the week before.

Dr. Neil Gehrels, the lead scientist for the Swift satellite, said flares of that magnitude could be expected just once in a millennium.

"That seems so improbable it's a puzzle right now," Dr. Gehrels said. "There's something going on here that we don't understand."

The radiation even temporarily compressed Earth's ionosphere, an envelope of charged gas at the top of the atmosphere, and distorted long-wavelength radio signals.

"It was really the big one," said Dr. Kevin Hurley, a researcher at the Space Sciences Laboratory at the University of California, Berkeley. "You could not have missed it."

Dr. Hurley and others pinpointed the origin of the pulse as a neutron star known as SGR 1806-20, about 50,000 light-years distant in the constellation Sagittarius. Neutron stars are remnants of stars after they have exploded in supernovas, and SGR 1806-20 is one of about 10 unusual neutron stars known as magnetars, which have extraordinarily strong magnetic fields, a quadrillion times as strong as Earth's.

Sudden shifts in the intense magnetic fields are believed to generate flares, in much the way the Sun generates solar flares, and two giant magnetar flares had been previously observed, one in 1979 and one in 1998. The Dec. 27 flare, however, was 100 times as powerful.

One physicist, Dr. David Eichler of Ben-Gurion University in Israel, wrote a paper in 2002 suggesting that magnetars might be capable of such cataclysmic flares, but most scientists, Dr. Gaensler said, "had no idea they could make a flare this big."

SGR 1806-20 has almost as much mass as 1.5 Suns, compressed into an incredibly dense ball about a dozen miles wide, and it spins around once every 7.5 seconds, slow for a neutron star. Discovered in 1979, SGR 1806-20 has at times been noisy, firing off small gamma ray flares, and at other times quiet. Its activity picked up in the past year.

"In retrospect, I guess you could say it was getting ready to let go," Dr. Hurley said, adding that he thought the magnetic fields, held in place by the crust of the star, had become twisted, building stress. "At some point, it gives way like an earthquake," he said.

Astronomers presented their observations at a NASA news conference on Friday, and several scientific papers describing the event will be published in a coming issue of the journal Nature.

In the aftermath of the flare, Dr. Gaensler, lead author of one of the Nature papers, and his colleagues used radio telescopes on Earth to track a shock wave radiating from the star. They were surprised to find the wave expanding rapidly, at a quarter of the speed of light, "which is not what you tend to see in the galaxy every day," he said.

SGR 1806-20 itself continues to spin as before, one revolution every 7.5 seconds. "Amazingly, the neutron star is still there," Dr. Gaensler said. "It did not explode or blow itself apart to bits."

The magnetar flare may help solve a cosmic mystery of gamma ray bursts, the prime mission of the Swift satellite. Bursts lasting from several seconds to a couple of minutes are believed to be caused by the collisions of black holes—events that are even more violent than magnetar flares and occur much farther away—but astronomers had been at a loss to explain shorter bursts lasting a couple of seconds or less.

Now they have at least a partial answer: some of the bursts are magnetar flares originating in other galaxies.

"It is clear magnetar flares make short gamma ray bursts," said Dr. Robert C. Duncan of the University of Texas. "It is at least a significant fraction of them."

But that still may not be the whole answer. Dr. Gehrels, the lead scientist for the Swift satellite, said that when astronomers looked in the direction of three recent short gamma ray bursts, those parts of the sky turned out to be empty.

"It just all fits so well, and then there were no galaxies there," Dr. Gehrels said.

But he said that as the satellite observed more gamma ray bursts, "we should know in the next couple of months the answer to this."

IN REVIEW

1. Discuss the event that happened in December 2004 that is described in the article. What mysteries are associated with this event? How has technology made the discovery of this event possible?
2. Describe the characteristics of magnetars. How do they differ from "regular" neutron stars?
3. What might have happened on magnetar SGR 1806-20 that resulted in an intense gamma ray burst? What surprising things were observed with this burst?
4. What uncertainties remain about the nature and origin of gamma ray bursts?
5. In your opinion, is it worthwhile to have orbiting observatories that can observe light wavelengths blocked by the Earth's atmosphere? In what ways, if any, have these orbiting observatories benefited humanity? In what ways, if any, have they been a burden on humanity? Discuss the benefits vs. the cost of space-based astronomy.

When a massive star explodes as a supernova, the core can collapse to form either a neutron star or a black hole. Now, evidence from the Chandra X-ray Observatory points to a third possibility: Some neutron stars may collapse further to create a quark star, with a density even greater than that of a neutron star but not quite high enough to force it to collapse to a black hole.

Stars Suggest a Quark Twist and a New Kind of Matter

By John Noble Wilford
***The New York Times,* April 11, 2002**

Observations of two stars, one unusually small and the other unusually cold, have led astronomers to think they are seeing evidence of a new form of matter and a new kind of star, one possibly made of elementary particles known as quarks and denser than any cosmic object other than a black hole.

As a group, quarks are building blocks of larger particles and are normally bound together as protons and neutrons. Now it appears that a certain combination of these quarks are capable of a more independent existence as "strange quark matter" inside quark stars.

Astronomers said yesterday that if their interpretation was correct, the new findings were startling and exciting because they promised to open a new window on the nature of matter on the tiniest scales.

The findings also seemed to challenge the standard model of neutron stars, until now the most extreme form of matter known to scientists.

Quarks have never been unambiguously observed outside a nucleus in Earth-bound laboratories, although theories of free quarks have been discussed for 20 years. Neutron stars are the compact remnants of dying massive stars that explode and collapse, and nothing denser had been observed.

A teaspoon of neutron star material weighs a billion tons, or as much as all the cars, trucks and buses on Earth. Matter in the suspected quark stars would be far denser.

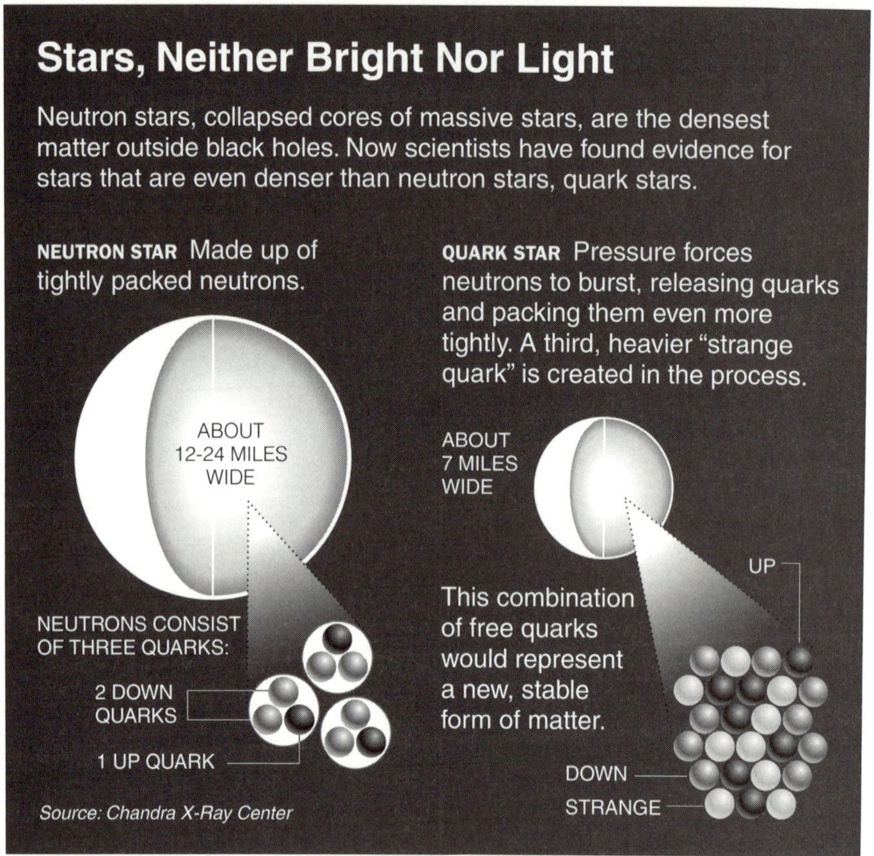

Bill Marsh/The New York Times

Physicists know of six kinds of quarks, two of which—up and down—make up ordinary matter like protons and neutrons. Theorists have long speculated that up and down quarks can be melded with a heavier kind known as the strange quark to form something called strange matter.

"The combined observational evidence points to a star composed not of neutrons, but of quarks in a form known as strange quark matter," said Dr. Jeremy Drake of the Harvard-Smithsonian Center for Astrophysics in Cambridge, Mass.

Dr. Sam Aronson, chairman of physics at Brookhaven National Laboratory on Long Island, where a high-energy particle collider is being used to study quarks, said the discovery, if confirmed, could lead to an understanding of the behavior of freely interacting quarks.

"No one has seen really free quarks," Dr. Aronson said.

Dr. Norman Glendenning, a senior scientist emeritus and quark specialist at Lawrence Berkeley National Laboratory in California, said at least one of the two stars could well be made only of quarks.

"If that is so," Dr. Glendenning said, "this star is in a class quite by itself and will be an astonishing discovery of fundamental significance."

But that does not mean the star is anomalous, other scientists said. If there are two quark stars, they said, such objects may turn out to be fairly common end-products of many exploded stars.

The research, described at a news conference at the National Aeronautics and Space Administration in Washington, was based primarily on data gathered by the Earth-orbiting Chandra X-ray Observatory. Dr. Drake is the lead author of a report to be published in June in The Astrophysical Journal.

Dr. Anne Kinney, director of astronomy and physics at NASA, cautioned, "I'd like to emphasize that this is evidence for, not proof of, a new form of matter."

The Chandra X-ray telescope, with help from the Hubble Space Telescope, studied two objects that were assumed to be neutron stars.

One is designated RXJ 1856, in the constellation Corona Australis, about 400 light-years from Earth. The other object is 3C58, in the constellation Cassiopeia and about 10,000 light-years away.

Analyzing the X-ray data, scientists found RXJ 1856 to be too small to be a neutron star. It is only seven miles in diameter, about half the diameter of a neutron star.

"We need a new theory to give us such a small star," Dr. Drake said.

Scientists also failed to detect the expected X-radiation from the hot surface of 3C58, indicating that its temperature must be far below that predicted for a neutron star.

The research on 3C58 was conducted by Dr. Patrick Slane and Dr. Steven Murray, both of the Center for Astrophysics, and Dr. David Helfand of Columbia University.

"Our observations offer the first compelling test of models for how neutron stars cool, and the standard theory fails," Dr. Helfand said. "It appears that neutron stars aren't pure neutrons after all. New forms of matter are required."

Each team qualified its conclusions.

Dr. Drake said his team's X-ray observations could be misleading if they were actually seeing a "hot spot" on a neutron star, making it only appear to be a small surface. The probability of this was considered small.

A stellar explosion observed by the Chinese and Japanese in 1181 presumably created 3C58, and this is the basis for calculating the object's predicted cooling rate.

The coordinates for the remnant from that explosion are not known precisely. If the temperature estimates actually apply to another object, this part of the research would be undermined.

But if it is the right object, scientists said, even a neutron star's density would not be enough to produce cooling particles fast enough to reduce its temperature to such a low level in the time since 1181. The object 3C58, Dr. Helfand said, would have to be as much as five times as dense for this to occur, indicating that at least the core of the object is made of something other than neutrons, perhaps "a new kind of exotic material."

Neutron stars, which have been known and observed for more than three decades, are formed out of the deaths of stars several times more massive than the Sun. After the star explodes, the core collapses to extremely high densities, which could be the force liberating the quarks.

"A neutron star, because it is so dense, may be the only natural place in the universe where quark matter exists," Dr. Glendenning said. "We may have discovered a way of learning if the existence of free quarks is true."

Dr. Michael Turner, a theoretical astrophysicist at the University of Chicago, suggested that the core collapse of an exploded star could at first produce a hot neutron star. Then, in a few seconds to a few days, the collapse could continue, producing even greater densities and thus freeing quark matter on its own.

If indeed there are quark stars, Dr. Turner said, they would probably be derived from collapsed stars that were 5 to 10 times the mass of the Sun.

The signatures of a quark star, Dr. Turner said, would seem to be its small size and rapid cooling.

"Regardless of how these mysteries are resolved, these precise observations are highly significant," Dr. Turner said. "They demonstrate our ability to use the universe as a laboratory where we can study some of the most fundamental questions of physics."

IN REVIEW

1. What are quarks?
2. Why is matter so dense in a neutron star? Why would it be more dense in a quark star?
3. Why would the existence of quark stars give us a new window into the nature of matter?
4. Describe the evidence from the Chandra X-ray Observatory that suggests the existence of quark stars.
5. How could the data from Chandra be misleading scientists? What other uncertainties remain about the existence of quark stars?

Our Milky Way galaxy holds stars of many ages, including some that are as old as almost any stars in the universe. It also hosts what seems to be a very massive black hole in its center. This article surveys recent discoveries about the nature of the galaxy that may shed light on the evolution of the universe as a whole.

In Galaxies Near and Far, New Views of the Universe Emerge

By John Noble Wilford
***The New York Times,* January 14, 2003**

Some astronomers came and exulted over glimpses of objects so far away and thus so long ago, about 13 billion light-years, that they opened eyes and minds to the universe as it was soon after stars and galaxies first began popping up everywhere.

They focused on three quasars, luminous objects thought to be powered by massive black holes, from that early time; one was the most distant quasar ever observed. They stretched their horizons at the sight of a new image from the Hubble Space Telescope showing a glittering expanse of galaxies so remote that some of them had not been seen before; a few may be the earliest in the universe.

Photo by Hubble Space Telescope/NASA

Acting as a cosmic lens in space, the gravity of a cluster of galaxies has amplified the light from galaxies near the dawn of time and smeared them into arcs in this picture taken by the Advanced Camera for Surveys on the Hubble Space Telescope and released last week. ABELL 1689, whose galaxies appear large and yellowish, is about 2.2 billion light-years away. DISTANT GALAXIES would be invisible without the amplification of the gravitational lens. ARCS AND DOTS: One galaxy, about 12 billion light-years away, has been split by the lens into two images. Many arcs, representing galaxies at a range of about 10 billion light-years, form a partial bull's-eye pattern around the center of the cluster. By analyzing these images, astronomers can map the mass, including unseen dark matter in the cluster.

For other scientists here at the winter conference of the American Astronomical Society, there was astonishment and mystery enough closer to home, in Earth's own Milky Way galaxy. X-ray emissions revealed the messy eating habits of the black hole at the galactic center, and radio waves showed the magnetic field there to be surprisingly tangled.

The newest and most impressive discovery was made on the fringes of the Milky Way. Two teams of astronomers have observed a previously unseen band of hundreds of millions of stars orbiting beyond the galaxy's main disk, more than twice as far from the galactic center as the solar system.

The most likely explanation, astronomers say, is that the ring stars are remnants of smaller galaxies that came too close and were captured by the Milky Way's overpowering gravity. This is more evidence, then, that in part the galaxy grew to its present size at the expense of less fortunate neighbors.

"This is a vivid smoking gun of the disruption of a satellite galaxy," said Dr. Bruce Margon of the Space Telescope Science Institute in Baltimore, who was not on either discovery team.

Part of the ring was observed by scientists from the Sloan Digital Sky Survey, a project for mapping one-quarter of the sky in three dimensions that is based at Sunspot, N.M. A team of Australian, British and Dutch astronomers, working at a telescope in the Canary Islands, then saw other sections, leading scientists to recognize that the band of stars appeared to reach around the entire galaxy.

The ring's encircling diameter is estimated at 120,000 light-years, the teams reported. Its thickness is about 10 times that of the rest of the galaxy, extending well above and below the galactic plane. Gravity, primarily from unseen dark matter of an unknown nature, holds the ring of up to 500 million stars—about the stellar population of small galaxies—in a nearly circular orbit.

Such a vast congregation of stars in a coherent ring had remained hidden from view because it lies in the same plane as the Milky Way disk and so was obscured by intervening stars, gas and dust. It was hard to distinguish the ring stars from the other matter and impossible to recognize their number or their organization in a discrete torus.

"Our entire picture of the Milky Way is being changed with this discovery," said Dr. Brian Yanny of the Fermi National Accelerator Laboratory outside Chicago, who was co-leader of the Sloan Survey group that detected the ring. Stars are more numerous and more closely spaced toward the center of galaxies and are expected to thin out more or less evenly in the outskirts.

The other leader, Dr. Heidi Jo Newberg of Rensselaer Polytechnic Institute in Troy, N.Y., said, "When we find large groups of stars formed into rings, it's an indication that at least part of our galaxy was formed by a lot of smaller or dwarf galaxies mixing together."

The Milky Way originated about 10 billion years ago, and it probably captured some neighbors several times since then, growing into a collection of about 400 billion visible stars. Other researchers at the meeting reported seeing a faint trail of stars that appear to be leftovers from a similar capture by Andromeda, a nearby galaxy not unlike the Milky Way.

But Dr. Annette Ferguson of the University of Groningen in the Netherlands, a member of the Canary Islands observing team, offered a possible alternative explanation for the ring. The stars there might have come from inside the galaxy's disc, she suggested, and gravitational interactions could have disturbed their orbits, causing them to migrate outward and into the ring.

In any case, scientists said, the ring promises to be an ideal place to study the mysterious dark matter, which makes up most of the mass of the universe, and its role in shaping cosmic structures.

Other researchers have been taking a closer, sharper look into the heart of the Milky Way, a region hidden from ordinary view by a fog of dust and gas. What they are finding in radio and X-ray observations surpasses immediate understanding.

The magnetic field near the galactic center, it seems, is more chaotic than previously thought.

Using the radio telescopes of the National Science Foundation's Very Large Array in New Mexico, a team of astronomers studied strange filaments produced by the interaction between the galaxy's magnetic field and high-velocity electrons. The filaments seen before were all aligned in nearly the same direction, like iron filings near a bar magnet.

But in the new observations, astronomers were surprised to see many more filaments oriented in various directions, tangled, they said, "like a bowl of spaghetti."

Dr. Joseph Lazio of the Naval Research Laboratory, a leader of the observing team, said, "The magnetic field in the center may not be all that organized or all that strong."

Another team leader, Dr. Namir Kassim, also of the Naval laboratory, could only conclude, "The Milky Way's center is an exciting, mysterious region that, once again, has given us a surprise."

And then there was the latest report on the eating habits of the black hole believed to lurk at the Milky Way's core.

A black hole, by definition, cannot be observed directly. It is an extremely compact massive object so dense that nothing, not even light, can escape its gravity. Scientists infer its presence by observing the turbulence and high-energy emissions its gravity creates on surrounding material.

Some black holes in other galaxies may weigh as much as several billion times the mass of the Sun, scientists think, but the Milky Way's is thought to be a puny three million times the solar mass.

In the longest and most sensitive X-ray observations made so far of the Milky Way's center, NASA's Earth-orbiting Chandra X-ray Observatory zoomed in on the region of the suspected black hole.

"We are getting a look at the everyday life of a black hole like never before," said Dr. Frederick K. Baganoff of the Massachusetts Institute of Technology. "We see it flaring on an almost daily basis."

What did this mean? The hot gases of these X-ray flares could be crumbs from the black hole's last meal of stellar matter, scientists said. Its table manners are atrocious.

The frequency of the X-ray flares suggests that the black hole eats often, scientists further said, but their weak intensity suggests that its meals are more like snacks than banquets.

"Although it appears to snack often, this black hole is definitely on a severe diet," Dr. Baganoff said. "This could be because explosive events in the past blew away much of the gas from the neighborhood of the black hole."

IN REVIEW

1. Describe each of the recent discoveries made about the Milky Way that are discussed in the article.
2. Explain how scientists discovered the ring of stars orbiting at the fringe of the Milky Way. Why had the ring been previously hidden?
3. How does the presence and structure of the ring give clues to how the Milky Way was formed? What uncertainties remain about the origin of the ring?
4. What are X-ray flares telling us about the black hole suspected to lie at the galactic center?
5. How do these new discoveries about our own galaxy bear on our understanding of the overall evolution of the universe?

In addition to the most famous space observatories such as the Hubble Space Telescope and Chandra X-ray Observatory, NASA and the space agencies of other nations have launched many smaller orbiting observatories. These smaller satellites have provided an incredible amount of data that have greatly increased scientific understanding of stars and galaxies. Once such small satellite, the Galaxy Evolution Explorer (Galex), was launched in April 2003. By detecting ultraviolet light, the spacecraft's mission is to observe hundreds of thousands of galaxies to learn more about the characteristics of galaxies and, specifically, the younger stars inside them.

Three Dozen New Galaxies Are Found in Nearby Space

By Dennis Overbye
***The New York Times,* December 22, 2004**

Fourteen billion years after the Big Bang started it all, there is still life in the old cosmos.

Astronomers announced yesterday that they had discovered three dozen baby galaxies in what passes for nearby space in the universe—two billion to four billion light-years distant. The galaxies, which are blossoming with new stars at a prodigious rate, resemble the infant Milky Way 10 billion years ago, the astronomers said.

Studying these new galaxies could give cosmologists new insights into the processes by which galaxies and stars first formed out of clouds of primordial gas and dust at the beginning of time.

"It's like looking out your window and seeing a dinosaur walk by," said Dr. Tim Heckman of Johns Hopkins University, who led a team using a NASA satellite, the Galaxy Evolution Explorer, or Galex, to pinpoint the newborns. Dr. Heckman spoke in Pasadena, Calif., at a news conference at the Jet Propulsion Laboratory, which manages the satellite. A paper describing the results has been submitted to The Astrophysical Journal.

The babies were a pleasant surprise.

Like the parents of a woman of a certain age who long ago gave up hope of grandchildren, astronomers had given up hope that the universe was still producing galaxies that could grow up to be the size of the Milky Way. The heyday of making stars, the active ingredients of galaxies, was five billion to eight billion years ago. Perhaps only dwarf galaxies were being born today.

"We didn't know if there were any newborns still around or if this phase of cosmic creation is over," Dr. Heckman explained.

The baby galaxies appear as bluish blobs of light about 10,000 light-years across in images sent back by the Galex satellite, which was launched in 2003 on a 29-month mission to survey the sky for ultraviolet emissions.

Ultraviolet light, which has a shorter wavelength than visible light, is produced by the hottest, most massive stars, like those of the Pleiades cluster, which shines in the sky above Orion these frigid crystalline nights. Because such stars do not last very long, they are also among the youngest stars in the sky.

As a result, young galaxies stand out in ultraviolet light, said Dr. Chris

Photo by National Optical Astronomy Observatory/NASA

The nearby galaxy Messier 81 as it appears in visible light and in ultraviolet light, which the NASA satellite Galaxy Evolution Explorer is using to map formation of galaxies across billions of years of cosmic time.

Martin of the Jet Propulsion Laboratory, the principal investigator for the Galex project. "Ultraviolet traces star formation," Dr. Martin said.

The hitch for astronomers hoping to study the recent evolution of stars and galaxies is that the atmosphere blocks ultraviolet rays from reaching Earth. So ultraviolet astronomy can be pursued only in space, with instruments like the Hubble Space Telescope and Galex.

Galex is designed to spot the ultraviolet glows of young stars and galaxies and thus help fill in the history of star formation and cosmic evolution over the last 10 billion years. It has a specially designed 20-inch-diameter telescope with a field of view four times as big as a full moon.

The new babies are only the first results of the project, and the astronomers said they expected to find more, although not many.

While they are not nearly the size of mature galaxies like the Milky Way, which is about 100,000 light-years across and has about 200 billion stars, the newborn galaxies outshine them in ultraviolet by a factor of 100 or so, which means they are producing stars "at a prodigious rate," in the words of Dr. Martin.

Dr. Alice Shapley, a theorist at the University of California, described them as "stragglers" of the great wave of galaxy formation that peaked when the universe was half its present age.

It is important, Dr. Shapley said, to try to find out what is finally causing these galaxies to form now. Are they accreting fresh star material from outside, for example? Indeed, she said, astronomers still do not know for sure whether these are really new galaxies, or whether perhaps they are old galaxies, hiding old stars inside them, that are undergoing a new burst of star formation.

These would be ideal objects to study with the Hubble Space Telescope, she added.

What will happen to these newborns is another mystery, Dr. Heckman said.

The infant Milky Way coalesced out of the murk 10 billion years ago, when the universe was more crowded and baby galaxies could bang into one another, merge and grow. "It's less clear what will happen in the future," Dr. Heckman said.

The universe is now a more diffuse place, and the baby galaxies may have been born into loneliness. If so, they will never grow up.

IN REVIEW

1. What are the characteristics of the infant galaxies discovered by Galex?
2. How far away are the newly discovered galaxies? How does their distance compare with the size of the observable universe? What does this suggest about how recently the galaxies were born? Can we be certain that they really are young galaxies?
3. In what way is the existence of these young galaxies surprising?
4. Why were these young galaxies easy to spot in ultraviolet light detected by Galex? Why is a satellite needed to detect UV light from the cosmos?
5. Why are these young galaxies rare in the universe? Explain why these small galaxies are not expected to grow to the size of our own galaxy, the Milky Way.

It has been more than 70 years since Edwin Hubble discovered the expansion of the universe. For most of the time since then, scientists assumed that the expansion should be slowing with time, as the mutual gravity of all the matter in the universe tries to pull itself back together. However, in the late 1990s, the observational data began to reveal a different scenario: Rather than slowing down, the expansion seems to be accelerating. New observations are gradually teaching us more about this apparent acceleration of the cosmos.

From Distant Galaxies, News of a "Stop-and-Go Universe"

By John Noble Wilford
***The New York Times,* June 3, 2003**

New observations of exploding stars far deeper in space, astronomers say, have produced strong evidence that the proportions of the mysterious forces dominating the universe have undergone radical change over cosmic history.

The findings, reported here at a meeting of the American Astronomical Society, which ended Thursday, supported the idea that once the universe was expanding at a decelerating rate but then began accelerating within the last seven billion years, scientists concluded.

NASA

A young supernova was recently detected in Arp 299, a colliding galaxy pair 140 million light-years away.

"We are now seeing hints that way back then the universe was slowing down," said Dr. John Tonry, an astronomer at the University of Hawaii who is a member of one team studying exploding stars, or supernovas, for signs of cosmic expansion rates.

The new research by Dr. Tonry's group and another, led by Dr. Saul Perlmutter of Lawrence Berkeley National Laboratory in California, confirmed the earlier surprising discovery that the universe is indeed expanding at an accelerating rate and has been for at least the last 1.2 billion years. But four supernovas, almost 7 billion light-years away, appeared to exist at a time the universe was slowing down, Dr. Tonry said.

"A stop-and-go universe" is the way Dr. Robert P. Kirshner of the Harvard-Smithsonian Center for Astrophysics characterized the phenomenon. Well, the expansion never really stopped, he conceded, but it has certainly revved up.

"Right now, the universe is speeding up, with galaxies zooming away from each other like Indy 500 racers hitting the gas when the green flag drops," said Dr. Kirshner, a member of the Tonry team. "But we suspect that it wasn't always this way."

The changing pace of cosmic expansion, combined with recently announced measurements of the cosmic microwave background, revealing conditions soon after the Big Bang, encourages theorists in thinking that a tug-of-war has been going on between dark forces of matter and energy no one yet understands.

The combined gravitational pull from all matter in the universe, most of which is beyond detection, has acted as a brake on cosmic expansion. The so-called dark matter apparently had the advantage when the universe was younger, smaller and denser. Now the ever-increasing pace of expansion suggests that something else even more mysterious is at work. Theorists are not sure what the antigravity force is, but they call it dark energy. It has apparently gained the upper hand.

This is the latest turn of events in the unfolding story of cosmic history. Once scientists believed the universe was everlastingly static. Along came Edwin P. Hubble, who discovered seven decades ago that the galaxies of stars are rushing away from one another in all directions. The universe, Hubble announced, is expanding.

Five years ago, astronomers were in for a surprise. They had assumed that after an initial burst of rapid expansion from the originating Big Bang the

Changes of Pace

Scientists now believe that after about 7 billion years of slowing down, dark energy overtook dark matter and the universe began to accelerate.

AGE OF UNIVERSE: 1 MILLION YEARS — 500 MILLION — 1 BILLION — 7 BILLION — EARTH, SUN BORN: 9.35 BILLION — 14 BILLION (TODAY)

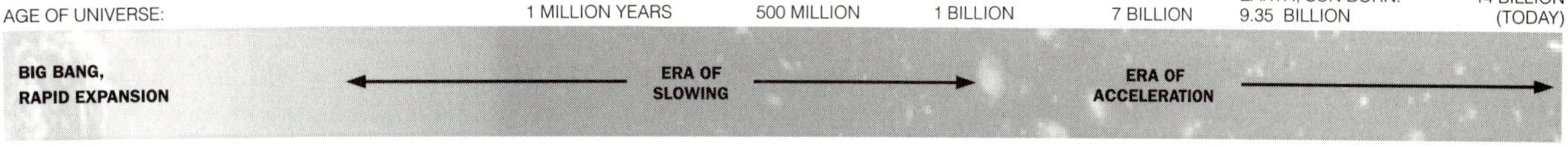

Source: Dr. Robert P. Kirshner, Harvard-Smithsonian Center for Astrophysics

The New York Times

gravity of matter was gradually slowing things down. Then the two supernova survey teams found that the universe was accelerating instead. This pointed to the existence of some kind of dark energy permeating all of space.

For the current research, astronomers observe what are called Type Ia supernovas, stellar explosions that at their peak are brighter than a billion stars like the Sun. They are thus visible across billions of light-years of space, and a close examination of their light reveals the distances, motions and other evidence of conditions. As the light travels to Earth, the wavelengths are stretched by an amount that reflects the universe's expansion when the star exploded.

Dr. Kirshner said the four extremely distant supernovas indicated that the universe seven billion years ago was "in fact winning this sort of cosmic tug-of-war," but now dark energy is more dominant.

Scientists said they assumed that with the stretching out of space the proportion of dark energy to dark matter had been reversed. In the earlier and denser universe, matter of all kinds, the invisible dark matter and the visible ordinary matter of stars and planets, predominated.

The team of Dr. Tonry and Dr. Kirshner estimates that about 60 percent of the universe is filled with dark energy and 30 percent of the mass is dark matter. The remaining 10 percent consists of ordinary matter, only 1 percent of which is visible in the galaxies. Theorists offer roughly the same estimates and surmise that the changeover from dark matter to dark energy domination probably occurred before 6.3 billion years ago.

Dr. Perlmutter said that much more research would be necessary to determine whether the changing density of the expanding universe was the only reason dark energy came to rule cosmic dynamics. Or have the physical properties of dark energy, whatever it is, changed?

Dr. Perlmutter said that in the words of Dr. Edward Witten, a theoretical astrophysicist at the Institute for Advanced Study at Princeton, the true nature of dark energy "would be No. 1 on my list of things to figure out."

The research teams are planning new observations of more distant supernovas to determine when cosmic acceleration began and to gather clues about the properties of dark energy. Some observations will be conducted with ground-based telescopes, others with the Hubble Space Telescope. Dr. Perlmutter's group has proposed putting a spacecraft in orbit with telescopes especially designed for supernova hunting and pinning down the nature of dark energy.

In "The Extravagant Universe," published last fall by Princeton University Press, Dr. Kirshner wrote: "We are not made of the type of particles that make up most of the matter in the universe, and we have no idea yet how to sense directly the dark energy that determines the fate of the universe. If Copernicus taught us the lesson that we are not at the center of things, our present picture of the universe rubs it in."

IN REVIEW

1. What evidence is being used in attempts to measure the changing rate of expansion of the universe?
2. Why was the discovery of acceleration surprising?
3. The title of the article refers to a "stop-and-go universe." What does this mean in the context of the article?
4. What evidence supports the idea of a "stop-and-go universe"?
5. What do we mean by dark energy, and what do scientists now know (or not know) about it?

In 1979, when Ridley Scott's film *Alien* was first released, the movie's poster said simply, "In space no one can hear you scream." This simple statement supports a basic property of physics that sound waves must travel through some type of medium (e.g., solid, liquid, or gas) and cannot travel through a vacuum. Although sound waves do travel through interstellar gas, the frequency of the sound is incredibly low due to the very low density of the medium. This fact, along with the incredibly low density of space, means that people could not hear any sounds in space. Nevertheless, sound waves traveling through interstellar space have recently been detected by the Chandra X-ray Observatory. Although Chandra detects light (X-rays specifically) and not sound, the wavelengths of the sound were inferred from different intensities of X-rays in a distant galactic cluster.

Songs of the Galaxies, and What They Mean

By Dennis Overbye
***The New York Times,* August 3, 2004**

Another black hole is singing, although it seems a little tone-deaf.

Last year, astronomers discovered that outbursts from a giant black hole were spreading pressure waves through the thin hot gas of the distant Perseus cluster of galaxies. The frequency of these waves was equivalent to a B flat, 57 octaves below middle C, the astronomers calculated.

Now another group of astronomers has discovered waves from another massive black hole spreading outward from the center of a galaxy known as M87 at the heart of the Virgo cluster, about 50 million light-years from Earth. They appear as rings and arcs of brightness in images of the galaxy obtained by NASA's Chandra X-Ray Observatory.

The sound waves in M87, spaced about four million years apart, are a little more than an octave higher than the Perseus black hole, they said, and a little rougher and less pure.

"If one could hear the sound, it would be more like the cannons in the '1812 Overture' than the pure tone of a musical instrument," said Dr. William Forman of the Harvard-Smithsonian Center for Astrophysics.

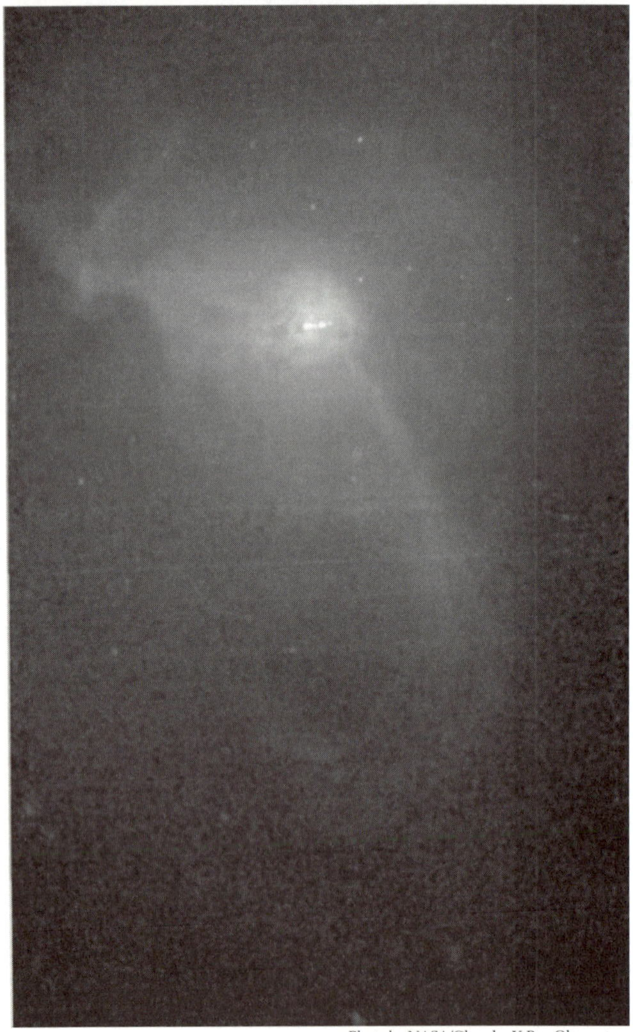

Photo by NASA/Chandra X-Ray Observatory

The galaxy M87, which is about 50 million light-years from Earth.

Frivolous as it may sound, the phenomenon is part of a changing picture of the relation between galaxies (and the other luminous structures that web the sky) and black holes, the Einsteinian monsters that seem to dwell in the heart of almost every galaxy.

In this view, black holes may regulate both their own growth and the growth of their galaxies by explosive outbursts, orchestrating the flow of gas that is essential to the formation of stars, the lifeblood of the luminous universe.

Dr. Abraham Loeb, a theorist at the Harvard-Smithsonian Center, said, "People have only recently come to realize that the central black holes affect the formation of galaxies in a profound way."

Because it is nearly next door in cosmic terms, M87 is a natural laboratory for examining how black holes relate to the galaxies around them.

"Now we see how the black hole grows, how it affects the gas from which it grows," Dr. Forman said. "It all seems to connect."

Dr. Forman is part of a multinational team of astronomers that has used Chandra and other

X-ray satellites to study what is going on in M87, and the lead author of a paper submitted to The Astrophysical Journal.

Speaking of the Virgo and Perseus results, Dr. Christine Jones of the Harvard-Smithsonian Center, also on the team, said, "We are seeing evidence of recurrent outbursts of the black holes in these galaxies."

It may seem contradictory to talk about a black hole releasing energy. According to Einstein's general theory of relativity, black holes are so dense that neither matter nor energy nor light nor anything else, including sound, can escape from them. They are sheer gravitational death. Nothing but inexorable doom, crushing and stretching, awaits anything that enters. A famous calculation by the Cambridge University theorist Dr. Stephen Hawking 30 years ago showed that when quantum effects are taken into account, black holes would leak and eventually explode, but the process would take much longer than the age of the universe, and so is little comfort.

But black holes can produce energy in the space around them by virtue of their enormous gravitational fields. Long before any infalling material, whether astronaut or gas cloud, reached the point of no return, it would be accelerated to nearly the speed of light and heated to millions of degrees, sparking X-rays and high-energy particles, as it swirled in a thick dense disk around the cosmic drain.

As a result of enormous pressure and magnetic fields, some of that stuff would never make it into the hole but would be squeezed like toothpaste and shot across space, away from the black hole in two oppositely directed jets.

The more black holes eat, the more they spill, and it is widely thought that their feeding frenzies power the violence seen in the nuclei of many galaxies, including the powerful quasars that are so bright they outshine their parent galaxies. Although the most massive black holes amount to only a tiny fraction of the mass of their galaxies, a tenth of a percent at most, they release huge quantities of energy, Dr. Loeb pointed out.

A wealth of observations of M87 over the years have illuminated, in messy, chaotic detail, the feeding habits of its central black hole, estimated to weigh in at two billion to three billion times the mass of the Sun. Dr. Forman said evidence of the messiness includes a sharp knotted spike, or jet, of energetic particles shooting from the galaxy's core; giant lobes of radio energy extending far out into intergalactic space; other bubbles and cavities, apparently inflated by the jet, in the inner regions of the galaxy; and a pair of arms of X-ray emission, one of which resembles the column of a rising mushroom cloud from a nuclear explosion.

The jet and an invisible counterpart going in the opposite direction, crashing into interstellar gas, seem to create the galaxy's discordant drum roll. In the new work, Dr. Forman and his collaborators combined two different Chandra pictures and data from the German Rosat and the European Space Agency's XMM-Newton satellites. The sound waves appear as bright arcs in the X-ray glow of hot gas that suffuses the M87 galaxy and the Virgo cluster in which it sits.

One of them makes an entire circle around the nucleus, at a radius of about 45,000 light-years. Another arc sits at a radius of about 57,000 light-years from the center. The most recent outburst, they estimated, was about 11 million years ago and was the equivalent of about 10 million supernova explosions.

The results echo the earlier ones from the Perseus cluster and strengthen the conclusion that through their outbursts black holes may be reaching across intergalactic space and controlling the fates of billions of stars not yet born.

Clusters of galaxies are the largest assemblages of matter in the universe, with masses up to a trillion Suns.

Most of the ordinary matter in them, however, is not in the form of stars but in a dilute gas that has been heated to millions of degrees by falling into the cluster. It has long been a puzzle why

Bubbling With Energy

A radio portrait of the galaxy M87 shows swirling fountains of energy that has been pumped out into the galaxy from the black hole at its center.

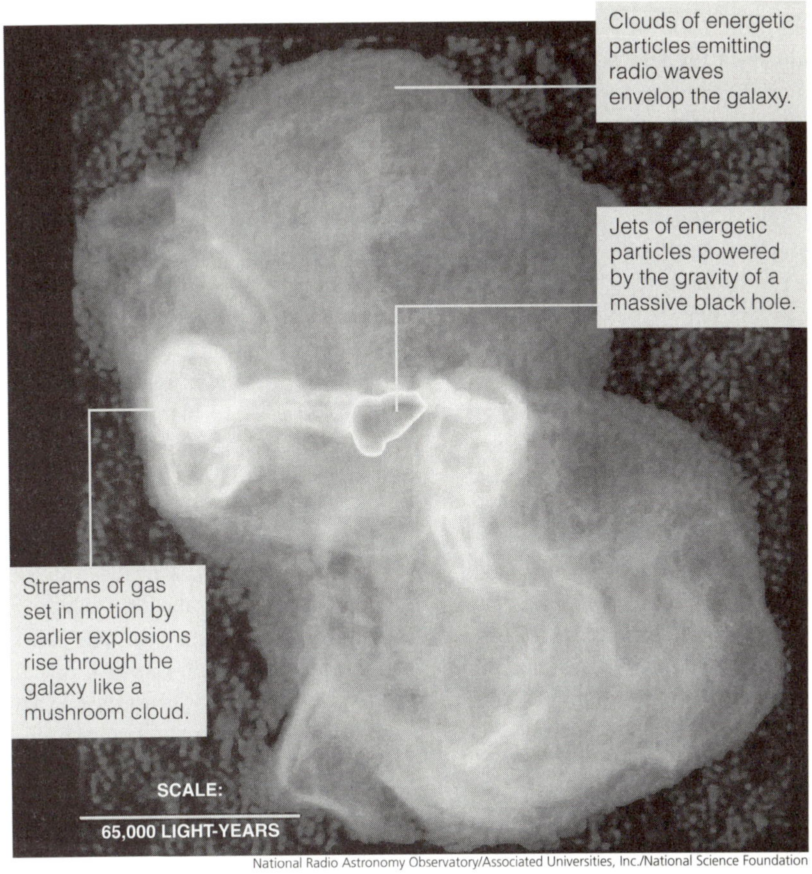

National Radio Astronomy Observatory/Associated Universities, Inc./National Science Foundation

this gas does not cool off, fall into the center of the cluster and make new stars.

Now it seems that singing black holes are heating the gas and keeping it from falling.

"This certainly shows that Perseus isn't an isolated case and shows that considerable energy is propagating out from the central engine into the intra-cluster gas," said Dr. Andrew Fabian of Cambridge University, who led the team that worked on Perseus.

As a result, the birth of new stars in those clusters has virtually halted. Dr. Peter Edmonds, another Harvard-Smithsonian astronomer who was not a member of either team, said in an e-mail message, "Billions or trillions of stars might be prevented from forming in M87 because of this black hole activity."

Such feedback activity from black holes, said Dr. Fabian and other astronomers, may be responsible for limiting the growth of galaxies. As the galaxy grows, the central black hole feeds and feeds from the accretion of gas, dust and smaller galaxies until its outbursts are strong enough to keep more material from falling into either the hole or the galaxy.

That kind of self-regulation, Dr. Loeb said, may explain a puzzle noted a few years ago. In 2000, work by two groups of astronomers, one led by Dr. Karl Gebhardt of the University of Texas, and another by Dr. Laura Ferrarese of Rutgers, found a mysterious correlation between the masses of the black holes and the masses of the galaxies in which they lived.

Bigger galaxies have larger gravitational fields and thus attract more material into their centers where the black holes feed, Dr. Loeb explained, and therefore black holes have to grow larger before their outbursts have enough oomph to hold off the gas falling into them.

But that view may be too simplistic, according to Dr. Anton Koekemoer of the Space Telescope Science Institute in Baltimore, who notes that the jets and outbursts from black holes actually seem to touch off the formation of stars by compressing interstellar gas along their boundaries, and to keep other gas from falling and condensing. "Both effects are going on at the same time," he said.

Just where these black holes came from is another problem.

Using Hubble as well as other telescopes, astronomers have inferred the presence of black holes, weighing up to a billion times the mass of the Sun, at the centers of most galaxies, including ones that existed only a billion years after the Big Bang itself. Earlier this year Stanford astronomers announced that they had found one with a mass of 10 billion times the mass of the Sun.

One theory is that these black holes are the remains of an early generation of very massive stars that condensed out of the primordial soup of hydrogen, 100 million to 200 million years after the Big Bang. They would rapidly burn out and explode, leaving behind a generation of "baby black holes," in Dr. Koekemoer's words, 100 or 200 times as massive as the Sun. These then grow by swallowing more gas or even each other. Dr. Loeb outlined an alternative scheme. Under certain circumstances, he explained, supermassive stars could be formed at the beginning. When they burned out and exploded these stars would leave behind black holes as much as 1,000 times as massive as the Sun.

"This would give you a jump-start," Dr. Loeb said.

Dr. Fabian said that how massive black holes could grow to large sizes without stifling their own growth was still an unsolved problem. Most of the growth of these black holes happened in the first few billion years of time, astronomers say, when the universe was dense with gas and the baby fragments of galaxies were furiously merging. The heyday of the quasars was about 10 billion years ago, when the universe was only four billion years old.

Since then the black holes of the universe have been relatively quiescent, including the one in our own Milky Way galaxy, only occasionally kicking up their heels in outbursts like the ones in M87 that are nonetheless and perhaps fortunately only a shadow of the energies of the quasars.

"A big black hole is kind of like a spider," Dr. Koekemoer said. "Most of the time it sits at the center of the galaxy doing nothing. Then a bug comes by. When it eats a bug, there is a burst of activity."

In the case of black holes, the bugs could be little galaxies, like the Milky Way's satellite, the Large Magellanic Cloud. And the result perhaps could be a chirp from our own resident black hole.

"If the Milky Way galaxy swallows the Large Magellanic Cloud, the black hole will light up," Dr. Koekemoer explained, adding, "It's been hungry a long time."

IN REVIEW

1. What are the frequencies of the sound waves discussed in the article? How do they compare with the frequencies of sounds that our ears can hear?
2. How do the sound waves give clues about how black holes grow and affect the galaxies in which they reside?
3. Explain how black holes release energy, even though nothing can escape from inside the black hole itself.
4. How does the release of energy from a supermassive black hole affect the galaxy in which it resides?
5. Describe a few of the major unanswered questions about supermassive black holes at the centers of galaxies.

Supermassive black holes are thought to reside at the centers of many galaxies. What would happen if a star passed too close to one? Recent observations may be telling us the answer.

Black Holes' Vast Power Is Documented

By John Noble Wilford
***The New York Times,* February 19, 2004**

New X-ray observations by orbiting satellites have given astronomers their first telling evidence that appears to confirm what had been only theory: that a star is doomed if it ventures too close to a supermassive black hole.

The National Aeronautics and Space Administration and the European Space Agency announced yesterday the detection of a brilliant flare of X-rays from the heart of a distant galaxy, followed by a fading afterglow.

After analysis, an international team of scientists concluded that the telescopes had witnessed the overpowering gravity of a black hole as it tore apart a star and gobbled up a hearty share of its gaseous mass.

It was an act of cosmic mayhem known as a stellar tidal disruption. It removed any lingering doubt, astronomers said, that the reputation of black holes as star-destroyers is fully deserved.

"Stars can survive being stretched a small amount, but this star was stretched beyond its breaking point," said Stefanie Komossa of the Max Planck Institute for Extraterrestrial Physics in Garching, Germany, who led the discovery team. "This unlucky star just wandered into the wrong neighborhood."

Astronomers said they suspected that the ill-fated star was thrown off course by a close encounter with another star. Then it fell under the gravitational influence of a black hole and its enormous tidal forces, nothing so benign as the Moon's tug on Earth's oceans. The black hole, in effect, reached out and squeezed and stretched the star until it disintegrated.

In a televised news briefing at NASA in Washington, G¸nther Hasinger, also an astrophysicist at the Max Planck Institute, said, "For the first time, we really are convinced that we are seeing a star being ripped apart by a black hole."

Alex Filippenko, an astronomy professor at the University of California, Berkeley, who was not involved in the research, agreed.

"This is really fantastic," Dr. Filippenko said. "This is very strong evidence that stars are being ripped apart by supermassive black holes."

The astronomers estimated that about 1 percent of the victimized star's mass was consumed by the black hole. This small amount, they said, was consistent with theoretical predictions that the momentum and energy of star-destruction process would cause most of the star's gas to be flung away from the black hole.

The initial outburst of high-energy radiation was detected in 1992 by the German Rosat X-ray spacecraft, but the observations were fuzzy. At the time, puzzled scientists could only speculate about the nature of the flare or its origin.

The more definitive observations were made three years ago by two orbiting X-ray telescopes, NASA's Chandra and the European XMM-Newton satellites. They determined that the X-rays were coming from the center of a galaxy, RX J1242-11. The most awesome black holes, with densely packed masses equivalent to millions or billions of Suns, are found at galactic cores. This one is estimated to have a mass of about 100 million Suns.

More detailed examinations of the energy spectrum of the X-rays by the European satellite, astronomers said, revealed physical conditions similar to the expected surroundings of black holes, ruling out other possible astronomical explanations. The energy liberated by the tidal disruption was reported to be equivalent to that of a supernova, an exploding star.

Dr. Filippenko said the findings should advance understanding of black holes and provide a critical framework for theoretical models of how they grow and evolve over time.

Black holes cannot be observed directly; their gravity is so strong that nothing, not even light, can escape their clutches. What is known about black holes is mostly deduced from observations of the whirlpool motions of gas and dust and stars under their gravitational influence.

The X-ray discovery may give scientists another means of identifying the presence of black holes and learning more about their behavior.

Kim Weaver, an astrophysicist at NASA's Goddard Space Flight Center in Greenbelt, Md., said the high-energy flares of a star's disintegration by black-hole gravity could serve as flashlights to illuminate the otherwise obscuring gas and dust and to provide glimpses into the inner regions of galaxies.

The astronomers gave assurances that the Sun is far enough away from a suspected black hole at the center of the Milky Way galaxy to be well out of danger.

IN REVIEW

1. What parts of the electromagnetic spectrum are used to study events taking place around black holes? Why do the observatories for these studies need to be located in space?
2. What are tidal forces? (Hint: See Chapter 5.) How would tidal forces affect a star that came close to a supermassive black hole?
3. What evidence suggests that we have observed the radiation produced by the destruction of a star near a supermassive black hole?
4. How do these observations help scientists understand how black holes evolve?

New observations of ripples moving through a galaxy cluster are shedding new light on how supermassive black holes can influence their surroundings out to very large distances. These ripples move like sound waves, and their nature provides the key clues.

Music of the Heavens Turns Out to Sound a Lot Like a B Flat

By Dennis Overbye
***The New York Times,* September 16, 2003**

Astronomers say they have heard the sound of a black hole singing. And what it is singing, and perhaps has been singing for more than two billion years, they say, is B flat—a B flat 57 octaves lower than middle C.

The "notes" appear as pressure waves roiling and spreading as a result of outbursts from a supermassive black hole through a hot thin gas that fills the Perseus cluster of galaxies, 250 million light-years distant. They are 30,000 light-years across and have a period of oscillation of 10 million years. By comparison, the deepest, lowest notes that humans can hear have a period of about one-twentieth of a second.

The black hole is playing "the lowest note in the universe," said Dr. Andrew Fabian, an X-ray astronomer at the Institute for Astronomy at Cambridge University in England.

Dr. Fabian was the leader on an international team that used NASA's Chandra X-ray Observatory to detect the black hole's notes as ripples of luminosity in the X-ray glow of the cluster. The discovery, announced last week at NASA headquarters in Washington and in a paper in the journal Monthly Notices of Royal Astronomical Society, might help solve longstanding problems regarding the structure of galaxy clusters, the largest, most massive objects in the universe, and the evolution of galaxies within them, astronomers said.

Far from being "just an interesting form of black hole acoustics," as Dr. Steven Allen of the Institute of Astronomy said in a news release, the sound waves might be the key to figuring out how such clusters grow.

Black holes, as decreed by Einstein's general theory of relativity, are objects so dense that neither light nor anything else, including sound, can escape them. But long before any sort of material disappeared into a black hole, theorists have surmised, it would be accelerated to near-light speeds by the hole's gravitational field and heated to millions of degrees as it swirled in a dense doughnut around the gates of doom, sparking X-rays and shock waves and squeezing jets of energy and particles across space.

Such furiously feeding black holes are thought to be the engines responsible for the violent quasars and other phenomena in the cores of galaxies. The new work suggests that such black holes can exert influences far beyond their host galaxies.

The biggest clusters, like the one in Perseus, can contain thousands of galaxies and trillions of stars. Paradoxically, most of the ordinary matter in them resides not in stars, but in intergalactic gas that has been heated by the fall into the cluster to temperatures of 50 million degrees or so. The gas glows brightly enough in X-rays to be seen far across the universe. Cosmologists use this X-ray glow to find clusters in the deep of spacetime.

It has long been a puzzle what keeps the cluster gas hot. Without a continuing input of energy, the gas at the center would radiate its heat, lowering its pressure, and cooler gas would flow in from the outskirts, providing fresh fuel to make stars.

"We should see stars forming in central galaxies," said Dr. Kimberly Weaver, an astronomer at the Goddard Space Flight Center in Greenbelt, Md. But they do not, she said.

Astronomers, Dr. Fabian said, suspected that black holes in the central galaxies of clusters might be keeping the cluster gas hot, but the astronomers did not know how.

As the brightest X-ray cluster in the sky, radiating 1,000 times as much energy in X-rays as visible light, Perseus is a logical laboratory for investigating the problem, Dr. Fabian explained. A particularly massive black hole is believed to lurk in a galaxy known as NGC 1275, which lies at the center of the cluster.

Two jets of radio energy shooting out of the galaxy's nucleus have blown two bubbles in the gas in the center of the cluster. In an X-ray image from the Chandra satellite released three years ago, these bubbles looked like the eyeholes of a giant eerie orange skull.

Last year, however, Dr. Fabian and his colleagues obtained a new long-exposure Chandra image of the Perseus cluster, which showed waves moving outward like ripples on a pond from the central bubbles.

The waves, they realized, might be the ideal missing link between the jets and the surrounding gas. Dr. Fabian compared the process to a child's

A supermassive black hole in the Perseus galaxy cluster is emitting sound waves 30,000 light-years across — a cosmic hum in B flat.

X-ray jets from the center of the hole create bubbles of high energy that push aside gases and make the giant waves.

X-ray view of the Perseus cluster (left), and enhanced to show sound waves.

Sources: Chandra X-Ray Center, Harvard-Smithsonian Center for Astrophysics; NASA

The New York Times; simulation image at top by Chandra X-Ray Center/A.Hobart; others by NASA

blowing bubbles in a glass of water through a straw. In this case, the jets are the straw. The bubbles pushing against the enormous pressure of the gas surrounding them create sound waves moving out through the cluster's gas, pumping energy into it and heating it.

Other astronomers called the results beautiful, but said more study was needed to confirm that the wave process could be supplying the missing energy to the cluster. "I think it might be," said Dr. Simon White of the Max Planck Institute for Astrophysics in Garching, Germany.

The energies are as prodigious as the symphony is boring. It takes the energy of 100 million supernova explosions to blow a central bubble in the cluster. If the black hole blows such bubbles continuously and it is this energy that has been keeping Perseus hot, then the black hole in Perseus must have been playing a steady B flat for a long time, said Dr. Fabian. "It's the longest-lasting symphony we know of," said Dr. Bruce Margon, an astronomer at the Space Telescope Science Institute.

IN REVIEW

1. How do the sound waves discussed in the article differ from those detectable by the human ear?
2. What is the Perseus cluster? How did the Chandra X-ray Observatory detect sound waves in the cluster?
3. How do scientists think the sound waves are linked to a supermassive black hole? Where is the black hole located?
4. How might the energy propagating through the cluster explain the X-ray glow from the cluster's gas?
5. How does understanding this cluster help scientists to understand the evolution of galaxies more generally?

As discussed in the book and in other articles, recent astronomical observations suggest that the rate of universal expansion of space is accelerating with time. The initial evidence for this acceleration came from supernova observations. The Chandra X-ray Observatory has now added another line of evidence pointing toward an accelerating universe.

By X-Raying Galaxies, Researchers Offer New Evidence of Rapidly Expanding Universe

By Dennis Overbye
***The New York Times,* May 19, 2004**

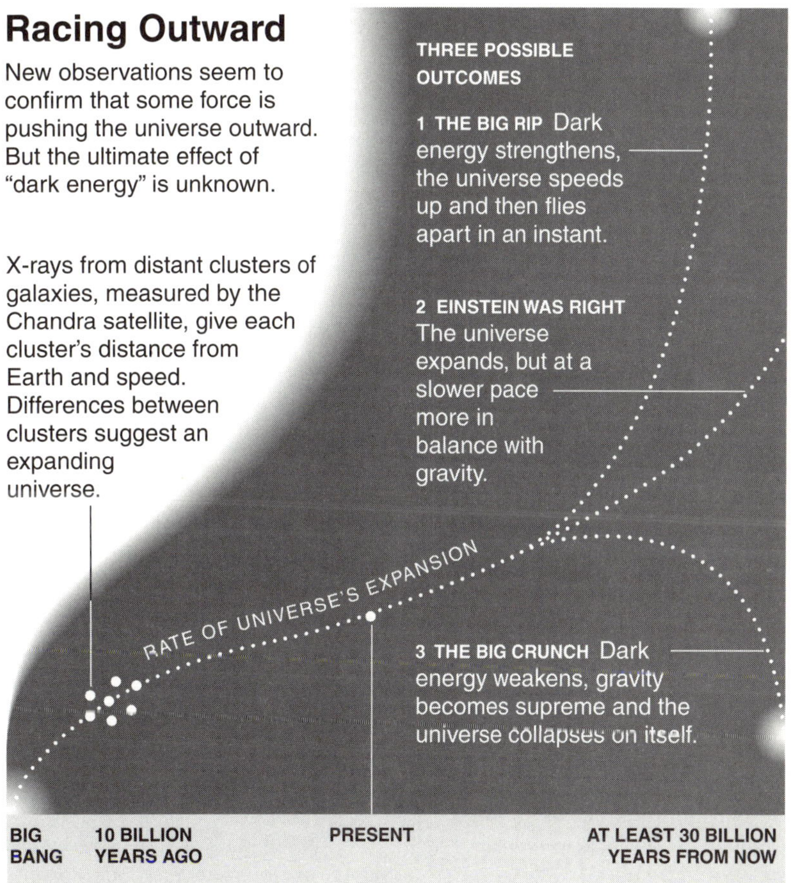

Source: Harvard-Smithsonian Center for Astrophysics — The New York Times

Correction Appended

Observations of giant clouds of galaxies far out in space and time have revealed new evidence that some mysterious force began to push the cosmos apart six billion years ago, astronomers said yesterday.

The results constitute striking confirmation of one of the weirdest discoveries of modern science: that the expansion of the universe seems to be accelerating, the galaxies flying apart faster and faster with time, under the influence of some antigravitational force. The work, astronomers said, opens up a powerful new way of investigating the nature of this "dark energy" and its effect on the destiny of the cosmos.

The astronomers used an orbiting X-ray satellite called Chandra to observe hot gases in the distant galactic clusters. By analyzing the X-rays emitted by those gases, they could calculate the distance from Earth and the speed of each of the clusters and thus trace the history of the expansion of the universe over the last 10 billion years, they said.

"The universe is accelerating," said Dr. Steve Allen of Cambridge University in England, leader of the international team that did the work. "We have found strong new evidence for dark energy."

They announced their results at a news conference at NASA headquarters in Washington. A paper describing the work has been submitted to the journal Monthly Notices of the Royal Astronomical Society.

Other astronomers hailed the X-ray cluster method as a potential complement to other ways of investigating dark energy but said they would withhold judgment about this particular calculation until they could study the details. Most of the previous studies, including those that led to the discovery of dark energy, used exploding stars known as Type 1a supernovas as cosmic distance markers.

Dr. Adam Riess of the Space Telescope Science Institute in Baltimore, an original discoverer of dark energy, hailed the work as another sign of the new age of "precision cosmology."

Dr. Riess said in an e-mail message: "Cosmologists are all from Missouri, the Show-Me State. It appears that X-ray clusters have been added as a new tool in our surveyor's tool kit. All tests point to a strange form of gravity we call dark energy. Some love it, some hate it; it appears we have to deal with it."

Dr. Martin Rees, a cosmologist at Cambridge who was not part of the team, called the results "neat work and a promising method," which, he noted, involved "very straightforward assumptions and simple physics."

Another cosmologist who was not part of the team, Dr. Michael Turner of the University of Chicago, said: "We can now be quite confident that the expansion of the universe is speeding up. It's not a fluke, it's not going away."

Dark energy has confounded experts and everybody else since two competing groups of astronomers discovered six years ago that the expansion of the universe was not slowing down due to cosmic gravity, as had been presumed, but was speeding up.

At the news conference, Dr. Andrew Fabian of Cambridge, a team member, compared the phenomenon to tossing an apple in the air and watching it go up faster and faster rather than falling back down. "It requires new physics beyond everyday experience, even the experience of an astronomer," Dr. Fabian said.

In recent years theorists have filled the journals with ever more fanciful explanations of what might be causing this behavior.

One possibility, first suggested and then rejected by Einstein, is that space itself has a repulsive force. But according to modern particle physics theory, this cosmological constant, as this force is known, should be about 1,060 times bigger than what astronomers have measured, causing theorists seek other explanations. Among them have been extra unseen dimensions to space, interactions with other, parallel universes and as-yet-undiscovered particles or forces.

Or perhaps, some theorists say, Einstein's theory of gravity, general relativity, which has been the backbone of cosmology for nearly a century, needs modification.

Astronomers hope that some answers will come if they can find out whether the density of dark energy—estimated to make up 75 percent of the universe—is changing with time.

If dark energy were constant, it would mean that Einstein's cosmological constant is in effect, and that most of the galaxies would move away too fast to be seen a mere 100 billion years from now.

If dark energy is increasing, it could mean the universe could end in a "big rip," in which even atoms would be torn apart. On the other hand, the dark energy could decrease and even turn into an attractive force, drawing the universe to an end in a "big crunch."

The new results are consistent with Einstein's cosmological constant but also allow for the possibility that the dark energy could be changing, echoing recent results from the supernova surveys.

"The nice thing is that this is a completely independent method based on very simple physics," Dr. Allen said. "It's the physics of hot gas and the physics of gravity."

Clusters of galaxies are the largest objects in the universe, containing thousands of galaxies and trillions of stars. But in a big cluster, the stars themselves are greatly outweighed by intergalactic gas, which has been condensed and heated to 100 million degrees or so by the cluster's immense gravity.

The X-rays that are spit out by this gas can be seen far across the universe. From their brightness astronomers can gauge the amount of gas in the cluster, and from the temperature of the gas, they can estimate the total mass in the galaxy cluster. Most of that mass is mysterious dark matter, which has been detected only by its gravitational effects on the luminous parts of the universe.

The astronomers made what they said was the simple assumption that clusters were a fair sample of the universe as a whole and that the cosmic ratio of dark matter to ordinary matter applied in each individual cluster. That allowed them to calculate distances to 26 clusters, from 1 billion to 10 billion light-years away, and thus measure how fast the universe was expanding when the light left those far-away galaxy clouds, confirming the cosmic acceleration.

"It's nice our results agree with previous experiments," Dr. Allen said. "It lets you feel rather more secure that everything is as it should be in those experiments."

Correction: May 21, 2004, Friday. Because of a transmission error, an article on Wednesday about a force that appears to be pushing the universe apart misstated the difference between the repulsive force as predicted by modern physics and the force actually measured. It is 1060 times as big, not 1,060 times.

IN REVIEW

1. How do observations made by the Chandra X-ray Observatory support the idea that universal expansion is accelerating?
2. Why are some astronomers reserving judgment about the observations made by Chandra?
3. Describe why dark energy is a frustrating concept for scientists. What are some of the possible explanations for dark energy?
4. How does the discussion on dark energy illustrate the hallmarks of science? In your opinion, what is the value of determining the true nature of dark energy?

How will the universe end? In simple terms, it seems that it should either continue to expand forever or someday stop expanding and begin to collapse. However, the idea of an accelerating expansion introduces new possibilities, including the possibility that the universe will eventually expand so rapidly that all matter will be torn apart in a "Big Rip."

From Space, a New View of Doomsday

By Dennis Overbye

***The New York Times,* February 17, 2004**

Once upon a time, if you wanted to talk about the end of the universe you had a choice, as Robert Frost put it, between fire and ice.

Either the universe would collapse under its own weight one day, in a fiery "big crunch," or the galaxies, now flying outward from each other, would go on coasting outward forever, forever slowing, but never stopping while the cosmos grew darker and darker, colder and colder, as the stars gradually burned out like tired bulbs.

Now there is the Big Rip.

Recent astronomical measurements, scientists say, cannot rule out the possibility that in a few billion years a mysterious force permeating space-time will be strong enough to blow everything apart, shred rocks, animals, molecules and finally even atoms in a last seemingly mad instant of cosmic self-abnegation.

"In some ways it sounds more like science fiction than fact," said Dr. Robert Caldwell, a Dartmouth physicist who described this apocalyptic possibility in a paper with Dr. Marc Kamionkowski and Dr. Nevin Weinberg, from the California Institute of Technology, last year.

The Big Rip is only one of a constellation of doomsday possibilities resulting from the discovery by two teams of astronomers six years ago that a mysterious force called dark energy seems to be wrenching the universe apart.

Instead of slowing down from cosmic gravity, as cosmologists had presumed for a century, the galaxies started speeding up about five billion years ago, like a driver hitting the gas pedal after passing a tollbooth.

Dark energy sounded crazy at the time, but in the intervening years a cascade of observations have strengthened the case that something truly weird is going on in the sky. It has a name, but that belies the fact that nobody really knows what dark energy is.

In six years it has become one of the central and apparently unavoidable features of the cosmos, the surprise question mark at the top of everybody's list, undermining what physicists presumed they understood about space, time, gravity and the future of the universe.

"In five years we've gone from saying it looks like a mistake to something that everyone is claiming evidence for," said Dr. Robert Kirshner of the Harvard-Smithsonian Center for Astrophysics, who was part of the original discovery.

Dr. Saul Perlmutter, a physicist from the Lawrence Berkeley Laboratory who was a leader of one of the 1998 teams, said he thought astronomers had even gotten comfortable with the idea—"or as comfortable as you can be with something as bizarre as dark energy."

Now, armies of astronomers are fanning out into the night, enlisting telescopes, large and small, from Chile to Hawaii to Arizona to outer space, in a quest to take the measure of dark energy by tracing the history of the universe with unprecedented precision.

Some of them are following the trail blazed by the first two groups six years ago, searching out a kind of exploding star known as Type 1a the supernova. Those stars serve as markers in space, enabling scientists to plumb the size of the universe and how it grew over time. Where the first groups based their conclusions on observing a few dozen supernovas, the new efforts intend to harvest hundreds or thousands of them.

Others are seeking to gain leverage by investigating how the antigravitational force of dark energy has retarded the growth of conglomerations of matter like galaxies. In one ambitious project, a team led by Dr. John Carlstrom of the University of Chicago is building an array of radio telescopes at the South Pole to count and study clusters of galaxies deep in space-time. Others are already probing the internal dynamics of galaxies by the thousands, or building giant cameras that use the light-bending powers of gravity itself as lenses to map invisible dark matter in space and compile a growth chart of cosmic structures.

Dr. Anthony Tyson, now at Bell Laboratories, is head of a project to build a "dark matter telescope" known as the Large Synoptic Survey Telescope. "High energy physicists have been marching into our project," he said. "This is not just another telescope. It's a physics experiment, like a particle accelerator."

After all, the fate of the universe is at stake. If the dark energy is virulent enough, then that fate "is quite fantastic and completely different than the possibilities previously discussed," Dr. Caldwell and his colleagues wrote last year.

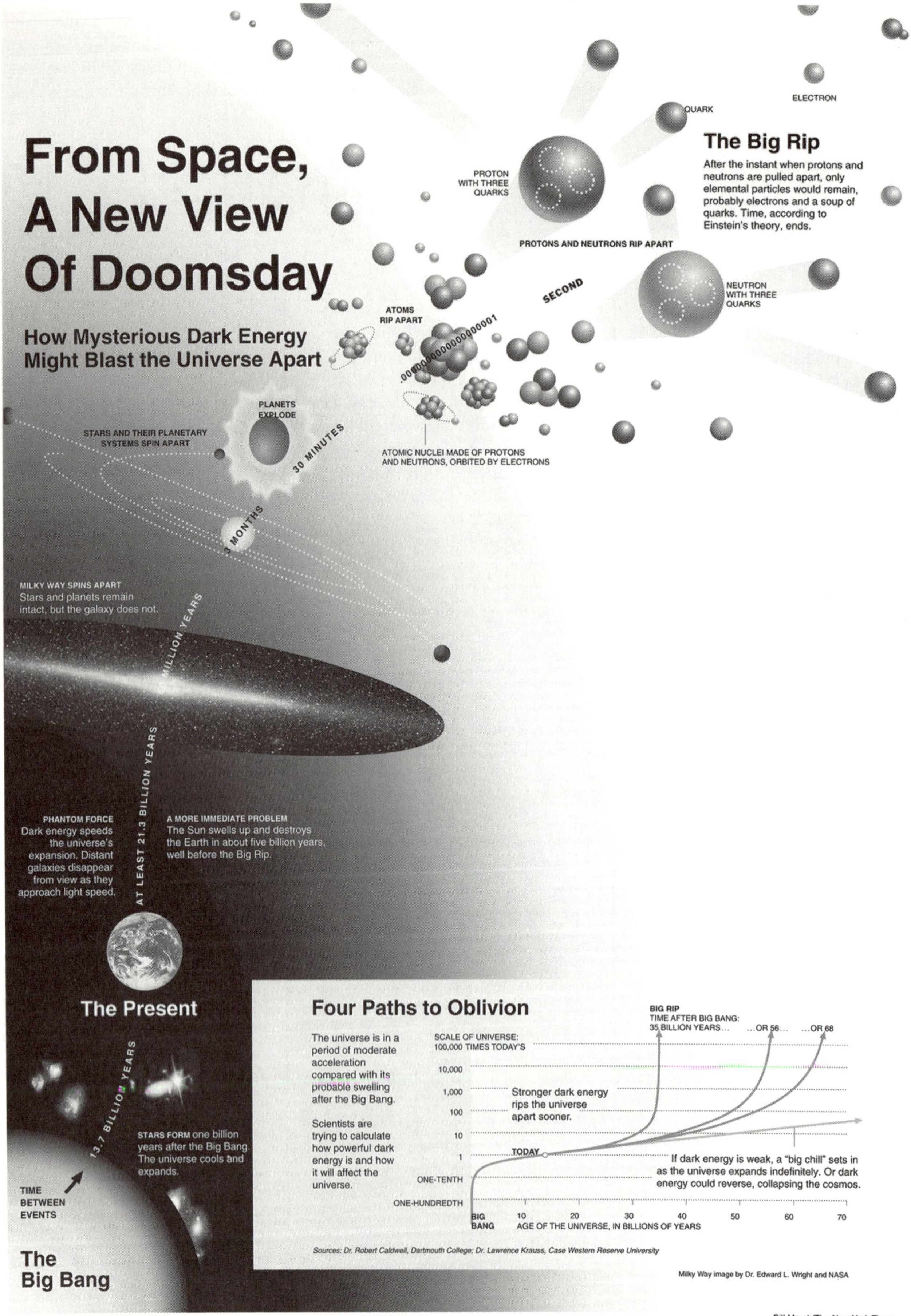

Bill Marsh/The New York Times

The Search for One Number

The idea of an antigravitational force pervading the cosmos does sound like science fiction, but theorists have long known that certain energy fields would exert negative pressure that would in turn, according to Einstein's equations, produce negative gravity. Indeed, some kind of brief and violent antigravitational boost, called inflation, is thought by theorists to have fueled the Big Bang.

As they try to figure out how this strange behavior could be happening to the universe today, astronomers say the ultimate prize from all the new observing projects could be as simple as a single number.

That number, known as w, is the ratio between the pressure and density of dark energy. Knowing this number and how it changes with time—if it does—might help scientists pick through different explanations of dark energy and thus the future of the universe—"whether it's gonna lead to a Big Rip, a Big Collapse or just a Big Fizzle," as Dr. Adam Riess of the Space Telescope Science Institute in Baltimore put it in an e-mail message.

One possible explanation for dark energy, perhaps the sentimental favorite among astronomers, is a force known as the cosmological constant, caused by the energy residing in empty space. It was first postulated back by Einstein in 1917. A universe under its influence would accelerate forever.

While the density of energy in space would remain the same over the eons, as the universe grows there would be more space and thus more repulsion. Within a few billion years, most galaxies would be moving away from our own faster than the speed of light and so would disappear from the sky; the edge of the observable universe would shrink around our descendants like a black hole.

But attempts to calculate the cosmological constant using the most high-powered modern theories of gravity and particle physics result in numbers 1060 times as great as the dark energy astronomers have observed—big enough, in fact, to have blown the universe apart in the first second, long before even atoms had time for form. Theorists admit they are at a loss. Perhaps, some of them now say, Einstein's theory of gravity, the general theory of relativity, needs to be modified.

Photo by Rick Friedman for The New York Times

Dr. Robert P. Kirshner of the Harvard-Smithsonian Center for Astrophysics suggests that jealousy, not gravity, is the strongest force.

Another possibility comes from string theory, the putative theory of everything, which allows that space could be laced with other energy fields, associated with particles or forces as yet undiscovered. Those fields, collectively called quintessence, could have an antigravity effect. Quintessence could change with time—for example, getting weaker and eventually disappearing as the universe expanded and diluted the field—or could even change from a repulsive force to an attractive one, which could set off a big crunch.

Recently, in a variation on the quintessence idea, Dr. Leonard Parker of the University of Wisconsin at Milwaukee and various colleagues, including Dr. Caldwell, have suggested that the field associated with some unknown very light particle could get tangled up with gravity and cause the universe to accelerate. That would alter Einstein's equations, said Dr. Caldwell. He added, "Our calculations show, however, that galaxies reside in a bubble of old-fashioned Einstein gravity, whereas gravity has changed outside and between galaxies."

A Weird Idea Gets Weirder

But the strangest notion is what Dr. Caldwell has called phantom energy, the dark energy that could lead to the Big Rip.

"It's weird negative pressure," said Dr. Lawrence M. Krauss, an astrophysicist at Case Western Reserve University in Cleveland.

While the density of the energy in Einstein's cosmological constant stays the same as the universe expands, the density of phantom energy would go up and up, eventually becoming infinite. Such would be the case if the parameter w turned out to be less than minus 1, say physicists, who admit they are stunned by the possibility and until recently simply refused to consider it.

"It crosses a boundary of good taste," Dr. Caldwell said, calling phantom energy "bad news stuff." Phantom energy violates physicists' intuitions about how the universe should behave. A chunk of it could be used to prop open wormholes in space and time—and thus create time machines, for example.

Photo by Peter DaSilva for The New York Times

Astrophysicist Saul Permuted sits by results from a earlier supernova that give scientists information that the universe is continuing to expand.

"It could lead to such bizarre effects as negative kinetic energy," Dr. Krauss said. As a result, objects like atoms would be able to lose energy by speeding up.

Nevertheless, a recent analysis by Dr. Caldwell and his Dartmouth colleague Dr. Michael Doran of the supernova measurements to date, combined with other cosmological data, suggest that w could lie anywhere from minus 0.8 to minus 1.25, leaving open the possibility of phantom energy. The cosmological constant would give a value of minus 1.0, and anything higher would be a sign of quintessence.

Dr. Kirshner said phantom energy had been dismissed as "too strange" when his group was doing calculations of dark energy back in 1998. In retrospect, he said, that was not the right thing to do.

"It sounds wacky," he said, referring to phantom energy, "but I think we're in a situation where we're going to need a really new idea. We're in trouble; the way out is going to be new imaginative things. It might be our ideas are not wild enough, they don't question fundamentals enough."

Dr. Chris Pritchet of the University of Victoria, who is part of a collaboration using the Canada-France-Hawaii telescope on Mauna Kea to search for supernovas, said, "In many ways phantom energy is unphysical, but we're not ruling it out."

Counting Down to the Big Rip

This version of doomsday would start slowly. Then, billions of years from now, as phantom energy increased its push and the cosmic expansion accelerated, more and more galaxies would start to disappear from the sky as their speeds reached the speed of light.

But things would not stop there. Some billions of years from now, depending on the exact value of w, the phantom force from the phantom energy will be enough to overcome gravity and break up clusters of galaxies. That will happen about a billion years before the Big Rip itself.

After that the apocalypse speeds up. About 900 million years later, about 60 million years before the end, our own Milky Way galaxy will be torn apart. Three months before the rip, the solar system will fly apart. The Earth will explode when there is half an hour left on the cosmic clock.

The last item on Dr. Caldwell's doomsday agenda is the dissolution of atoms, 10-19, a tenth of a billionth of a billionth of a second before the Big Rip ends everything.

"After the rip is like before the Big Bang," Dr. Caldwell said. "General relativity says: "The end. Time can't evolve."

The cosmos probably still has a lot of life in it, according to recent calculations by Dr. Krauss. Based on the current age of the universe, some 14 billion years, and other data, w cannot be less than about minus 1.2, he said, putting the Big Rip about 55 billion years in the future.

"It can't be very phantom," Dr. Krauss said.

The dark energy surveys now under way hope to be able to measure w to an accuracy of 5 percent, but even if that can be done, it may not be sufficient to eliminate the nightmare of phantom energy.

"It's hard to measure anything in astronomy to a few percent," said Dr. Sandra Faber of the University of California at Santa Cruz, who directs one of the dark energy surveys. Variations in the atmosphere and gaps in astronomers' understanding of supernova explosions add uncertainty to the dark energy measurements.

As a result some astronomers fear that the results may leave us on the razor's edge unable to decide between a cosmological constant and the other possibilities—quintessence or a Big Rip. Cosmologists could then be stuck with a "standard model" of the universe that fits all the data, but which they have no hope of understanding.

If the parameter w comes out to be something other than minus 1, Dr. Krauss said, it will at least give some direction to physicists.

One encouraging sign—"a tantalizing bit of hope," in Dr. Krauss's words—that the data will distinguish between a cosmological constant and the other possibilities came last fall when Dr. Riess, of Baltimore, reported, based on new observations of distant supernovas, that the "cosmic jerk" when dark energy took over the universe happened only five billion years ago.

In the standard cosmological constant model, said Dr. Riess, the turnaround should have come one or two billion years earlier.

Dr. Tyson was more sanguine. "Dark energy is crazy, right?" he said. "It's going to be exciting no matter what we find."

Dark Future for Dark Energy?

The work of the dark energy hunters has been complicated by the impending loss of the Hubble Space Telescope, which can see far enough out in space and time to measure how and if the dark energy parameter w is changing over the eons.

Last month, citing safety, NASA canceled all future shuttle maintenance missions to the telescope, dooming it to die in orbit, probably within three years, according to astronomers. "The Hubble shutdown will slow us all down," Dr. Perlmutter said.

At the same time, as a result of the agency's presidentially ordered shift toward the Moon and Mars, plans for a special satellite that was to have been jointly sponsored by NASA and the Department of Energy have at least temporarily disappeared from NASA's five-year budget plan.

Dr. Perlmutter, who has devoted much of his time in the last six years to the proposed satellite experiment, said he hoped that a way would be found to keep the project on track to be launched in the next decade.

"When you have the most exciting scientific problem of the day, you don't want to wait around," he said.

IN REVIEW

1. Briefly describe the possible fate of the universe that scientists are calling the Big Rip.
2. Where does the idea of the Big Rip come from? What other possible fates for the universe are consistent with current data?
3. What do scientists mean by a "cosmological constant"? How is it related to "dark energy"?
4. Why are scientists having such a difficult time explaining the observed acceleration of the cosmos?
5. In your opinion, how close are we to knowing the eventual fate of the universe? Do you think we will ever know it with confidence? Defend your opinion.

The prior article discussed the possibility that matter throughout the universe will eventually be torn apart in a "Big Rip." This article discusses additional data shedding light on possible fates of the universe.

New Data on 2 Doomsday Ideas, Big Rip vs. Big Crunch

By James Glanz
***The New York Times,* February 21, 2004**

A dark unseen energy is steadily pushing the universe apart, just as Einstein predicted, suggesting the universe may have a more peaceful end than recent theories envision, according to striking new measurements of distant exploding stars by the orbiting Hubble Space Telescope.

The energy, whose source remains unknown, was named the cosmological constant by Einstein. In a prediction in 1917 that he later called "my greatest blunder," Einstein posited a kind of antigravity force that was pushing galaxies apart with a strength that did not change over billions of years of cosmic history.

Theorists seeking to explain the mysterious force have suggested that it could, in fact, become stronger or weaker over time—either finally tearing the universe apart in a violent event called "the big rip" or shutting down in the distant future—tens of billions of years from now.

If the force somehow shut down, gravity would again predominate in the cosmos, and the universe would collapse on itself. That version of oblivion is sometimes called "the big crunch."

On Friday, Dr. Adam Riess of the Space Telescope Science Institute in Baltimore presented the first broad set of observational figures that gauge the strength of the antigravity force over time. The information, he said, suggests that the cosmos will gradually expand, cool and darken, more akin to a slide into senescence rather than a violent apocalypse.

Dr. Riess and his team, which included Drs. Louis Strolger of the science institute and Alexei V. Filippenko of the University of California at Berkeley, used the Hubble to search for exploding stars, or supernovae, that are swept up in the cosmic push of the dark energy. They discovered 42 new supernovas in their survey area, including six of the seven most distant known.

Photo by Marty Katz for The New York Times

Dr. Adam Riess of the Space Telescope Science Institute says his data on antigravity suggest the universe will slowly expand, cool and darken.

Rather than seeing the changes in the push that many theorists had predicted, Einstein's unchanging cosmological constant fits the data better than any of the alternatives.

"What we've found is that it looks like a semipermanent kind of dark energy," Dr. Riess said. "It appears like it's been with us for a long time. If it is changing, it's doing so slowly. Einstein's theory is looking a lot better than before this data."

A cosmologist not involved in the work, Dr. Michael S. Turner of the University of Chicago, said: "This is the biggest mystery in all of science, whether or not dark energy varies with time. It's a big, big clue, and this is the first information we have."

Although the new results favor Einstein's nearly century-old prediction, Dr. Turner said, they still do not rule out some alternative theories. The information specifically leaves open the chance that the antigravity force will eventually strengthen and tear apart planets, stars and even atoms in a big rip, among other exotic possibilities.

Dr. Turner said that future measurements were quite likely to turn up smaller changes in the force over time and that those subtleties could help unravel the mystery.

The results do suggest that any ultimate cataclysm could not occur for, perhaps, 30 billion years, Dr. Riess said. Several physicists said that those estimates were highly uncertain, but that the findings could lead to a wide-ranging reassessment of models that predict strange variable energy densities in space.

"Models which predict wild dark energy densities which change a lot with time don't look so good," Dr. Yun Wang, a cosmologist at the University of Oklahoma, said. "Everybody will go back to the drawing board."

Dr. Filippenko presented some of the results at a conference here on Sources and Detection of Dark Matter and Dark Energy in the Universe.

The measurements raise new questions about the NASA decision, which is under review, to let the Hubble die a slow death in space rather than try another service mission with a space shuttle. The administrator of National Aeronautics and Space Administration, Sean O'Keefe, has said a service mission would be too risky after the disaster of the Columbia space shuttle.

Dr. Riess said he disagreed with the decision to stop the Hubble, which would halt the research for years to come. He pointed out that the remarkable clarity of the work depended on the Advanced Camera for Surveys, which spacewalking astronauts installed in Hubble two years ago.

The organizer of the conference here, Dr. David B. Cline of the University of California, Los Angeles, said that in view of the team's results, he was inclined to agree with Dr. Riess's assessment.

"They really shocked everybody by showing they could do this," Dr. Cline said. "You have to say obviously it's shame that the Hubble can't continue its life."

All the scientists agreed that because the problem had become so knotty, a complete solution might have to wait for a proposed satellite, tentatively called the Supernova/Acceleration Probe, or SNAP, which is planned to observe thousands of exploding stars from space.

The End of It All

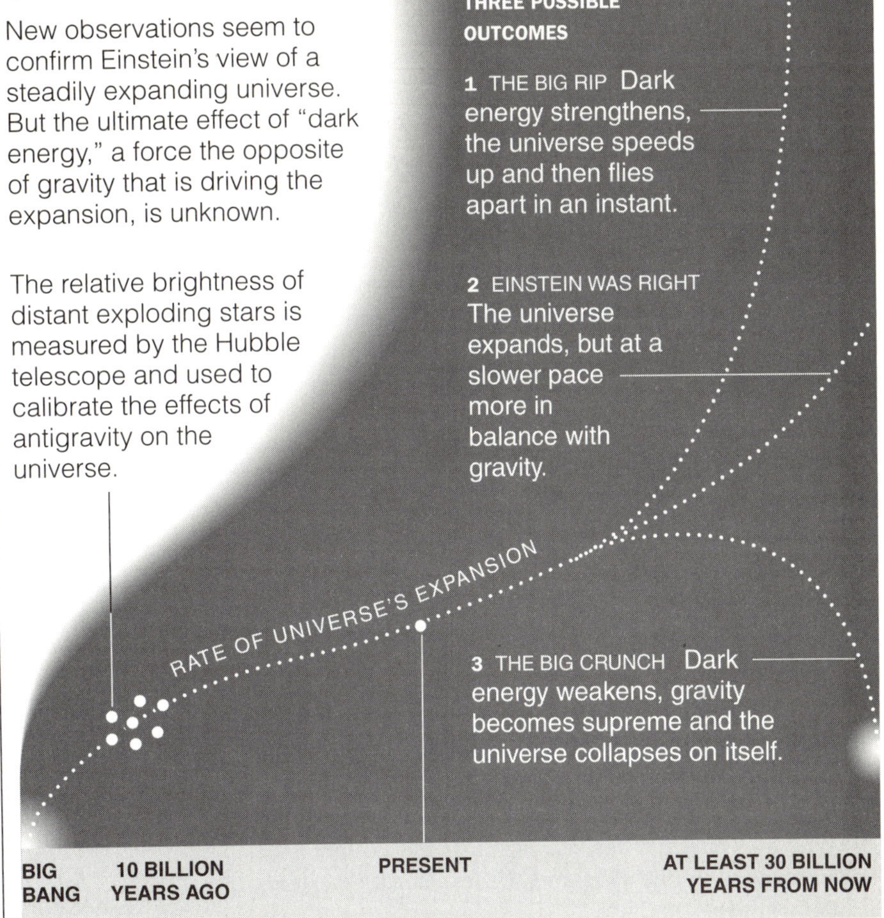

Source: Dr. Adam Riess, Space Telescope Science Institute — The New York Times

"It's so tantalizing and so beautiful to see this kind of data set," said Dr. Saul Perlmutter, a physicist at Lawrence Berkeley National Laboratory who is leading the SNAP team. "This kind of result is so exciting for those of us who are eager to get to that next step."

Einstein invented the cosmological constant 1917 to explain why the universe, filled with countless stars and galaxies that are attracted to one another by gravity, did not collapse on itself. He added a constant term, called lambda, to his equations of general relativity, which describe the workings of gravity and the curvature of space on large scales. Lambda provided a repulsive force to counteract gravity.

In 1929, the astronomer Edwin Hubble, for whom the space telescope is named, discovered that the universe was expanding—from the primeval explosion called the Big Bang, scientists would learn. In that picture, gravity would slow the expansion and, possibly, reverse it at a far-off time. Einstein abandoned lambda as unnecessary and called it a blunder.

Six years ago, however, two supernova groups—one led by Dr. Perlmutter and the other by several scientists, including Drs. Filippenko and Riess—found that cosmic expansion was speeding up rather than slowing. The evidence was that distant supernovas, swept up in some sort of cosmic repulsion, were farther away than they would otherwise be and therefore appeared dimmer.

Lambda was reborn, but astronomers still had little information on whether the antigravity force was truly constant. The latest results suggest, within the accuracy of the measurements, at least, that something like Einstein's cosmological constant is indeed at work.

"His greatest blunders are our greatest ideas," Dr. Sean M. Carroll, a cosmologist at the University of Chicago, said. "It is a triumph of general relativity."

Einstein's theory says nothing about why the energy of the cosmological constant, or anything else, should be filling space in the first place, Dr. Carroll said. Theorists have turned to explanations as bizarre as parallel universes that exert an influence on our own or extra unseen dimensions in the fabric of space.

Dr. Carroll said the supernova findings might raise the stock of certain ideas that have emerged from string theory, which assumes that fundamental physics is a manifestation of minuscule strings that vibrate in an 11-dimensional space. String theorists have calculated what they call cosmic "landscapes," in which empty space contains energy densities that resemble a cosmological constant.

Because dark energy has never been directly detected and all these explanations are so bizarre, a few scientists say the ultimate explanation might be something else entirely.

Others conceded that there was a bit of a letdown that nearly a century after Einstein invented the cosmological constant nary a crack in his theory had turned up.

"This gives me a sinking feeling," Dr. Max Tegmark of the University of Pennsylvania said. "My nightmare is that we're going to be forever stuck with this puzzle."

IN REVIEW

1. What evidence supports the idea that some sort of "dark energy" is acting in the universe? What does dark energy do? How does its existence affect our ideas about the fate of the universe?
2. Why did Einstein introduce the cosmological constant, lambda, into his general theory of relativity? Why did he later call his decision to include it the biggest blunder of his career? Why might it have been?
3. Why are scientists now taking Einstein's cosmological constant seriously? How might it explain the acceleration of the cosmos?
4. The article alludes to the fact that the prior NASA administrator cancelled all future plans for servicing the Hubble Space Telescope. How would the loss of the Hubble Space Telescope affect efforts to understand the acceleration of the cosmos?
5. In mid-2005, the new NASA administrator announced that NASA is taking a new look at servicing the Hubble Space Telescope. What is the current status of plans for servicing it?

The cosmic microwave background shows slight temperature variations that are thought to reveal the density variations in the infant universe that ultimately led to the pattern of galaxy clusters and voids observed today. This article discusses another line of evidence supporting current ideas about how our current lumpy universe took shape. According to scientific theory, extremely small-scale disturbances, called quantum fluctuations, caused matter to clump together. These clumps, in turn, generated sound waves, with matter clustering at the ripples. With observations from telescopes in New Mexico and Australia, scientists confirmed this theory, and gained a better understanding of the characteristics of the matter that makes up the universe.

Vestiges of Big Bang Waves Are Reported

By Kenneth Chang
***The New York Times*, January 12, 2005**

Astronomers reported on Tuesday that they had convincingly seen, in the patterns of galaxies scattered across the night sky, the vestiges of sound waves that rumbled through the universe after the Big Bang.

Stars and galaxies tended to form along the ripples of the sound waves where matter was slightly denser, and the pull of gravity was slightly stronger. The ripples preserve a picture of the universe when it was only about one million years old and fit well with astronomers' ideas of how the universe, which started smooth and uniform, became lumpy with stars, gas clouds and other celestial objects.

Two teams of researchers analyzing the locations of thousands of galaxies from two sections of the sky reported similar findings on the sound waves at a meeting of the American Astronomical Society here.

Earlier research had found signs of the ripples, but "we regard this as smoking-gun evidence," said Dr. Daniel Eisenstein of the University of Arizona, lead investigator of one of the teams.

"The important picture we have of the universe is hanging together amazingly well," said Dr. Martin Rees, a professor of cosmology and astrophysics at Cambridge University, who was not involved with either team. "The standard picture is firming up."

According to that picture, matter was evenly distributed in all directions for the first instant after the Big Bang. But then burbling caused by the physics of quantum mechanics created slight imperfections, clumps that were slightly denser with ordinary matter, as well as dark matter, the unknown material that accounts for most of the mass in the universe.

Just as ripples spread out from a pebble dropped in a pond, sound waves spread out from the dense clumps, traveling about half the speed of light through the hot gas made of matter, which is composed of electrons and protons, and of photons, or particles of light.

About 400,000 years after the Big Bang, the universe cooled enough that the charged electrons and protons combined to form hydrogen atoms, which allowed most of the photons to escape the hot gas. Several years ago, astronomers detected the sound waves etched by the photons.

The sound waves continued to spread for an additional 600,000 years, and when the last remaining photons escaped, the waves stopped, roughly 500,000 light-years from the dense clumps that produced them. When stars began to form, they tended to form around either the pebble-like clumps of dark matter or along the ripples.

As the universe has expanded in the 13.7 billion years since then, the typical distance between ripple and clump has stretched to 500 million light-years.

The new research shows the matter component of the early sound waves. Galaxies in the present universe are more likely to be 500 million light-years apart than other distances, Dr. Eisenstein said. One light-year is the distance light travels in one year, or 6 trillion miles.

The pictures do not show sharply delineated ripples, because the ripples were small and many overlapping ripples emanated from many different clumps.

"It's a much more subtle effect than that," Dr. Eisenstein said. "It's like you've taken a handful of gravel and thrown them in a pond."

Dr. Eisenstein and his colleagues used information from the Sloan Digital Sky Survey, which is mapping galaxies with a telescope in New Mexico.

The other team used data from a project called the 2dF Galaxy Redshift Survey that is scanning the sky with a telescope in Australia.

The research has also refined estimates on the amount of matter—18 percent of matter is ordinary matter that makes up stars and planets, and the remaining 82 percent is dark matter. And it offers further evidence that the geometry of the universe is perfectly flat, where the angles of all triangles always add up to 180 degrees.

"It's more than confirmation of what we already knew from the microwave background," said Dr. Richard S. Ellis, a professor of astronomy at the California Institute of Technology and a member of the 2dF team.

As the astronomers look farther away and further back in time, the size of the ripples will decrease in size. The ripples could serve as a convenient yardstick to track the history of the universe's expansion. That could shed light on dark energy, a mysterious force discovered in the past few years that, at cosmological distances, is stronger than gravity and is causing the expansion of the universe to accelerate.

IN REVIEW

1. According to current models, how did sound waves travel through the early universe and help clump matter together?
2. How do patterns in the distribution of galaxies seem to be revealing the sound waves of the early universe?
3. Based on the new observations, how does the amount of ordinary matter in the universe compare with the amount of dark matter? What is the significance of this result?
4. What do astronomers mean by a "flat" universe? What is the significance of the new evidence suggesting that the universe is "flat"?
5. What cosmic riddles could the sound waves help solve? Explain.

Physicists have long thought that nature should be inherently simple. The current picture, in which four forces govern interactions among many types of particles, seems too complex. About 30 years ago, scientists successfully showed that two of the four forces—the electromagnetic and weak forces—merge together into a single electroweak force at high energies. But the "holy grail" of physics is finding a single theory that explains all four forces and the diversity of particles. String theory attempts to do this, and progress in string theory is summarized in this article.

String Theory, at 20, Explains It All (or Not)

By Dennis Overbye
***The New York Times,* December 7, 2004**

They all laughed 20 years ago.

It was then that a physicist named John Schwarz jumped up on the stage during a cabaret at the physics center here and began babbling about having discovered a theory that could explain everything. By prearrangement men in white suits swooped in and carried away Dr. Schwarz, then a little-known researcher at the California Institute of Technology.

Only a few of the laughing audience members knew that Dr. Schwarz was not entirely joking. He and his collaborator, Dr. Michael Green, now at Cambridge University, had just finished a calculation that would change the way physics was done. They had shown that it was possible for the first time to write down a single equation that could explain all the laws of physics, all the forces of nature—the proverbial "theory of everything" that could be written on a T-shirt.

And so emerged into the limelight a strange new concept of nature, called string theory, so named because it depicts the basic constituents of the universe as tiny wriggling strings, not point particles.

"That was our first public announcement," Dr. Schwarz said recently.

By uniting all the forces, string theory had the potential of achieving the goal that Einstein sought without success for half his life and that has embodied the dreams of every physicist since then. If true, it could be used like a searchlight to illuminate some of the deepest mysteries physicists can imagine, like the origin of space and time in the Big Bang and the putative death of space and time at the infinitely dense centers of black holes.

In the last 20 years, string theory has become a major branch of physics. Physicists and mathematicians conversant in strings are courted and recruited like star quarterbacks by universities eager to establish their research credentials. String theory has been celebrated and explained in best-selling books like "The Elegant Universe," by Dr. Brian Greene, a physicist at Columbia University, and even on popular television shows.

Last summer in Aspen, Dr. Schwarz and Dr. Green (of Cambridge) cut a cake decorated with "20th Anniversary of the First Revolution Started in Aspen," as they and other theorists celebrated the anniversary of their big breakthrough. But even as they ate cake and drank wine, the string theorists admitted that after 20 years, they still did not know how to test string theory, or even what it meant.

As a result, the goal of explaining all the features of the modern world is as far away as ever, they say. And some physicists outside the string theory camp are growing restive. At another meeting, at the Aspen Institute for Humanities, only a few days before the string commemoration, Dr. Lawrence Krauss, a cosmologist at Case Western Reserve University in Cleveland, called string theory "a colossal failure."

©2004 Rick Friedman

Cambridge, MA, Nov. 5, 2004. Harvard University Department of Physics. Andrew Strominger Professor of Physics, LEFT, and Cumrun Vafa. Microscopic origin of the Bekenstein Hawking Entropy. They are holding what they believe could be a representation of the string in string theory.

Photo by Laura Pedrick

Princeton, NJ, Oct. 31, 2004. Dr. Ed Whitten is a leading theoretical physicist on the subject of string theory. Here, he continues his work on the campus of the Institute for Advanced Study.

String theorists agree that it has been a long, strange trip, but they still have faith that they will complete the journey.

"Twenty years ago no one would have correctly predicted how string theory has since developed," said Dr. Andrew Strominger of Harvard. "There is disappointment that despite all our efforts, experimental verification or disproof still seems far away. On the other hand, the depth and beauty of the subject, and the way it has reached out, influenced and connected other areas of physics and mathematics, is beyond the wildest imaginations of 20 years ago."

In a way, the story of string theory and of the physicists who have followed its siren song for two decades is like a novel that begins with the classic "what if?"

What if the basic constituents of nature and matter were not little points, as had been presumed since the time of the Greeks? What if the seeds of reality were rather teeny tiny wiggly little bits of string? And what appear to be different particles like electrons and quarks merely correspond to different ways for the strings to vibrate, different notes on God's guitar?

It sounds simple, but that small change led physicists into a mathematical labyrinth, in which they describe themselves as wandering, "exploring almost like experimentalists," in the words of Dr. David Gross of the Kavli Institute for Theoretical Physics in Santa Barbara, Calif.

String theory, the Italian physicist Dr. Daniele Amati once said, was a piece of 21st-century physics that had fallen by accident into the 20th century.

And, so the joke went, would require 22nd-century mathematics to solve.

Dr. Edward Witten of the Institute for Advanced Study in Princeton, N.J., described it this way: "String theory is not like anything else ever discovered. It is an incredible panoply of ideas about math and physics, so vast, so rich you could say almost anything about it."

The string revolution had its roots in a quixotic effort in the 1970's to understand the so-called "strong" force that binds quarks into particles like protons and neutrons. Why were individual quarks never seen in nature? Perhaps because they were on the ends of strings, said physicists, following up on work by Dr. Gabriele Veneziano of CERN, the European research consortium.

That would explain why you cannot have a single quark—you cannot have a string with only one end. Strings seduced many physicists with their mathematical elegance, but they had some problems, like requiring 26 dimensions and a plethora of mysterious particles that did not seem to have anything to do with quarks or the strong force.

When accelerator experiments supported an alternative theory of quark behavior known as quantum chromodynamics, most physicists consigned strings to the dustbin of history.

But some theorists thought the mathematics of strings was too beautiful to die.

In 1974 Dr. Schwarz and Dr. Joel Scherk from the École Normale Supérieure in France noticed that one of the mysterious particles predicted by string theory had the properties predicted for the graviton, the particle that would be responsible for transmitting gravity in a quantum theory of gravity, if such a theory existed.

Without even trying, they realized, string theory had crossed the biggest gulf in physics. Physicists had been stuck for decades trying to reconcile the quirky rules known as quantum mechanics, which govern atomic behavior, with Einstein's general theory of relativity, which describes how gravity shapes the cosmos.

That meant that if string theory was right, it was not just a theory of the strong force; it was a theory of all forces.

"I was immediately convinced this was worth devoting my life to," Dr. Schwarz recalled "It's been my life work ever since."

It was another 10 years before Dr. Schwarz and Dr. Green (Dr. Scherk died in 1980) finally hit pay dirt. They showed that it was possible to write down a string theory of everything that was not only mathematically consistent but also free of certain absurdities, like the violation of cause and effect, that had plagued earlier quantum gravity calculations.

In the summer and fall of 1984, as word of the achievement spread, physicists around the world left what they were doing and stormed their blackboards, visions of the Einsteinian grail of a unified theory dancing in their heads.

"Although much work remains to be done there seem to be no insuperable obstacles to deriving all of known physics," one set of physicists, known as the Princeton string quartet, wrote about a particularly promising model known as heterotic strings. (The quartet consisted of Dr. Gross; Dr. Jeffrey Harvey and Dr. Emil Martinec, both at the University of Chicago; and Dr. Ryan M. Rohm, now at the University of North Carolina.)

The Music of Strings

String theory is certainly one of the most musical explanations ever offered for nature, but it is not for the untrained ear. For one thing, the modern version of the theory decreed that there are 10 dimensions of space and time.

To explain to ordinary mortals why the world appears to have only four dimensions—one of time and three of space—string theorists adopted a notion first bruited by the German mathematicians Theodor Kaluza and Oskar Klein in 1926. The extra six dimensions, they said, go around in sub-submicroscopic loops, so tiny that people cannot see them or store old National Geographics in them.

A simple example, the story goes, is a garden hose. Seen from afar, it is a simple line across the grass, but up close it has a circular cross section. An ant on the hose can go around it as well as travel along its length. To envision the world as seen by string theory, one only has to imagine a tiny, tiny six-dimensional ball at every point in space-time.

But that was only the beginning. In 1995, Dr. Witten showed that what had been five different versions of string theory seemed to be related. He argued that they were all different manifestations of a shadowy, as-yet-undefined entity he called "M theory," with "M" standing for mother, matrix, magic, mystery, membrane or even murky.

In M-theory, the universe has 11 dimensions—10 of space and one of time, and it consists not just of strings but also of more extended membranes of various dimension, known generically as "branes."

This new theory has liberated the imaginations of cosmologists. Our own universe, some theorists suggest, may be a four-dimensional brane floating in some higher-dimensional space, like a bubble in a fish tank, perhaps with other branes—parallel universes—nearby. Collisions or other interactions between the branes might have touched off the Big Bang that started our own cosmic clock ticking or could produce the dark energy that now seems to be accelerating the expansion of the universe, they say.

Photo by J. Emilio Flores/For the Times

Dr. Joseph Polchinski at the Institute for Theoretical Physics at University of California Santa Barbara.

Toting Up the Scorecard

One of string theory's biggest triumphs has come in the study of black holes. In Einstein's general relativity, these objects are bottomless pits in space-time, voraciously swallowing everything, even light, that gets too close, but in string theory they are a dense tangle of strings and membranes.

In a prodigious calculation in 1995, Dr. Strominger and Dr. Cumrun Vafa, both of Harvard, were able to calculate the information content of a black hole, matching a famous result obtained by Dr. Stephen Hawking of Cambridge University using more indirect means in 1973. Their calculation is viewed by many people as the most important result yet in string theory, Dr. Greene said.

Another success, Dr. Greene and others said, was the discovery that the shape, or topology, of space, is not fixed but can change, according to string theory. Space can even rip and tear.

But the scorecard is mixed when it comes to other areas of physics. So far, for example, string theory has had little to say about what might have happened at the instant of the Big Bang.

Moreover, the theory seems to have too many solutions. One of the biggest dreams that physicists had for the so-called theory of everything was that it would specify a unique prescription of nature, one in which God had no choice, as Einstein once put it, about details like the number of dimensions or the relative masses of elementary particles.

But recently theorists have estimated that there could be at least 10100 different solutions to the string equations, corresponding to different ways of fold ing up the extra dimensions and filling them with fields—gazillions of different possible universes.

Some theorists, including Dr. Witten, hold fast to the Einsteinian dream, hoping that a unique answer to the string equations will emerge when they finally figure out what all this 21st-century physics is trying to tell them about the world.

But that day is still far away.

"We don't know what the deep principle in string theory is," Dr. Witten said.

For most of the 20th century, progress in particle physics was driven by the search for symmetries—patterns or relationships that remain the same when we swap left for right, travel across the galaxy or imagine running time in reverse.

For years physicists have looked for the origins of string theory in some sort of deep and esoteric symmetry, but string theory has turned out to be weirder than that.

Recently it has painted a picture of nature as a kind of hologram. In the holographic images often seen on bank cards, the illusion of three dimensions is created on a two-dimensional surface. Likewise string theory suggests that in nature all the information about what is happening inside some volume of space is somehow encoded on its outer boundary, according to work by several theorists, including Dr. Juan Maldacena of the Institute for Advanced Study and Dr. Raphael Bousso of the University of California, Berkeley.

Just how and why a three-dimensional reality can spring from just two dimensions, or four dimensions can unfold from three, is as baffling to people like Dr. Witten as it probably is to someone reading about it in a newspaper.

In effect, as Dr. Witten put it, an extra dimension of space can mysteriously appear out of "nothing."

The lesson, he said, may be that time and space are only illusions or approximations, emerging somehow from something more primitive and fundamental about nature, the way protons and neutrons are built of quarks.

The real secret of string theory, he said, will probably not be new symmetries, but rather a novel prescription for constructing space-time.

"It's a new aspect of the theory," Dr. Witten said. "Whether we are getting closer to the deep principle, I don't know."

As he put it in a talk in October, "It's plausible that we will someday understand string theory."

Tangled in Strings

Critics of string theory, meanwhile, have been keeping their own scorecard. The most glaring omission is the lack of any experimental evidence for strings or even a single experimental prediction that could prove string theory wrong—the acid test of the scientific process.

Strings are generally presumed to be so small that "stringy" effects should show up only when particles are smashed together at prohibitive energies, roughly 1019 billion electron volts. That is orders of magnitude beyond the capability of any particle accelerator that will ever be built on earth. Dr. Harvey of Chicago said he sometimes woke up thinking, What am I doing spending my whole career on something that can't be tested experimentally?

This disparity between theoretical speculation and testable reality has led some critics to suggest that string theory is as much philosophy as science, and that it has diverted the attention and energy of a generation of physicists from other perhaps more worthy pursuits. Others say the theory itself is still too vague and that some promising ideas have not been proved rigorously enough yet.

Dr. Krauss said, "We bemoan the fact that Einstein spent the last 30 years of his life on a fruitless quest, but we think it's fine if a thousand theorists spend 30 years of their prime on the same quest."

The Other Quantum Gravity

String theory's biggest triumph is still its first one, unifying Einstein's lordly gravity that curves the cosmos and the quantum pinball game of chance that lives inside it.

"Whatever else it is or is not," Dr. Harvey said in Aspen, "string theory is a theory of quantum gravity that gives sensible answers."

That is no small success, but it may not be unique.

String theory has a host of lesser known rivals for the mantle of quantum gravity, in particular a concept called, loop quantum gravity, which arose from work by Dr. Abhay Ashtekar of Penn State and has been carried forward by Dr. Carlo Rovelli of the University of Marseille and Dr. Lee Smolin of the Perimeter Institute for Theoretical Physics in Waterloo, Ontario, among others.

Unlike string theory, loop gravity makes no pretensions toward being a theory of everything. It is only a theory of gravity, space and time, arising from the applications of quantum principles to the equations of Einstein's general relativity. The adherents of string theory and of loop gravity have a kind of Microsoft-Apple kind of rivalry, with the former garnering a vast majority of university jobs and publicity.

Dr. Witten said that string theory had a tendency to absorb the ideas of its critics and rivals. This could happen with loop gravity. Dr. Vafa; his Harvard colleagues, Dr. Sergei Gukov and Dr. Andrew Neitzke; and Dr. Robbert Dijkgraaf of the University of Amsterdam report in a recent paper that they have found a connection between simplified versions of string and loop gravity.

"If it exists," Dr. Vafa said of loop gravity, "it should be part of string theory."

Looking for a Cosmic Connection

Some theorists have bent their energies recently toward investigating models in which strings could make an observable mark on the sky or in experiments in particle accelerators.

"They all require us to be lucky," said Dr. Joe Polchinski of the Kavli Institute.

For example the thrashing about of strings in the early moments of time could leave fine lumps in a haze of radio waves filling the sky and thought to be the remains of the Big Bang. These might be detectable by the Planck satellite being built by the European Space Agency for a 2007 launching date, said Dr. Greene.

According to some models, Dr. Polchinski has suggested, some strings could be stretched from their normal submicroscopic lengths to become as big as galaxies or more during a brief cosmic spurt known as inflation, thought to have happened a fraction of a second after the universe was born.

If everything works out, he said, there will be loops of string in the sky as big as galaxies. Other strings could stretch all the way across the observable universe. The strings, under enormous tension and moving near the speed of light, would wiggle and snap, rippling space-time like a tablecloth with gravitational waves.

"It would be like a whip hundreds of light-years long," Dr. Polchinski said.

The signal from these snapping strings, if they exist, should be detectable by the Laser Interferometer Gravitational Wave Observatory, which began science observations two years ago, operated by a multinational collaboration led by Caltech and the Massachusetts Institute of Technology.

Another chance for a clue will come in 2007 when the Large Hadron Collider is turned on at CERN in Geneva and starts colliding protons with seven trillion volts of energy apiece. In one version of the theory—admittedly a long shot—such collisions could create black holes or particles disappearing into the hidden dimensions.

Everybody's favorite candidate for what the collider will find is a phenomenon called supersymmetry, which is crucial to string theory. It posits the existence of a whole set of ghostlike elementary particles yet to be discovered. Theorists say they have reason to believe that the lightest of these particles, which have fanciful names like photinos, squarks and selectrons, should have a mass-energy within the range of the collider.

String theory naturally incorporates supersymmetry, but so do many other theories. Its discovery would not clinch the case for strings, but even Dr. Krauss of Case Western admits that the existence of supersymmetry would be a boon for string theory.

And what if supersymmetric particles are not discovered at the new collider? Their absence would strain the faith, a bit, but few theorists say they would give up.

"It would certainly be a big blow to our chances of understanding string theory in the near future," Dr. Witten said.

Beginnings and Endings

At the end of the Aspen celebration talk turned to the prospect of verification of string theory. Summing up the long march toward acceptance of the theory, Dr. Stephen Shenker, a pioneer string theorist at Stanford, quoted Winston Churchill:

"This is not the end, not even the beginning of the end, but perhaps it is the end of the beginning."

Dr. Shenker said it would be great to find out that string theory was right.

From the audience Dr. Greene piped up, "Wouldn't it be great either way?"

"Are you kidding me, Brian?" Dr. Shenker responded. "How many years have you sweated on this?"

But if string theory is wrong, Dr. Greene argued, wouldn't it be good to know so physics could move on? "Don't you want to know?" he asked.

Dr. Shenker amended his remarks. "It would be great to have an answer," he said, adding, "It would be even better if it's the right one."

IN REVIEW

1. What is the goal of string theory? What is the basic idea behind it?
2. Describe a few of the successes of string theory over the past 20 years.
3. Why is it so difficult to put string theory to the test? Given that difficulty, is string theory really a "theory" by our usual scientific definition? Explain.
4. Why are some scientists skeptical of whether string theory is even on the right track, let alone of its being the right theory to explain the nature of the universe?
5. Based on what you've read, how much effort do you think scientists should continue to make in pursuing string theory? Defend your opinion.

The movie *Contact* opens with a scene in which the character played by Jodie Foster is shown patiently listening through headphones amidst giant radio telescopes. The character is not listening to music, but to static that may contain some faint signal from intelligent life beyond Earth. Although the movie is science fiction, it is based on actual work being done by the SETI Institute, a private, nonprofit organization whose mission is to promote deeper understanding of the potential for life beyond Earth. The SETI Institute is not focused entirely on searching for extraterrestrial intelligent life; institute scientists also study the origin of life on Earth and the potential for life of any kind on other worlds.

Search for Life Out There Gains Respect, Bit by Bit

By Dennis Overbye
***The New York Times*, July 8, 2003**

Years after Congress ordered NASA to pull the plug on a survey looking for alien radio signals from the stars, the Search for Extraterrestrial Intelligence, or SETI, as it is known to aficionados, seems to have gradually achieved a modicum of respect in the halls of Washington.

The most recent indication appeared at the end of last month, when NASA named 12 groups that had won five-year grants to participate as "lead teams" in its Astrobiology Institute, which investigates the origin and future of life in the universe.

On the list was the SETI Institute, an organization in Mountain View, Calif., that has carried on the abandoned survey.

The group proposed a variety of basic research on the way planetary environments affect life or are affected by it. One project is aimed at determining whether certain kinds of stars are promising abodes for life and thus good targets for a planned expansion of the institute's search for intelligent radio signals.

That would make the grant the first money in a decade that NASA has allocated for work related to radio searches, the astronomers at the institute said.

For most of the last decade, "SETI was a four-letter word in NASA," said Dr. Frank Drake, a radio astronomer and former chairman of the SETI Institute. "It was not uttered in speeches, or in documents."

NASA said nothing had changed. The agency does not as a rule finance ground-based astronomy and, thus, has no SETI (pronounced SEH-tee) program. But SETI research can be supported as long as it meets the strictures of good science and emerges from a competitive peer-reviewed process, explained Dr. Edward J. Weiler, associate administrator of the NASA Office of Space Science.

Dr. Michael Meyer, a biologist who heads the Astrobiology Institute, described the proposed study as "pure astronomy," aimed at looking for potential habitable planets, research that fits with the institute's mission. NASA, he added, is eager to use the results to find targets for its planned Terrestrial Planet Finder satellite.

Astronomers around SETI and elsewhere said NASA and Congress had recently shown warming attitudes toward the politically embattled subject of intelligent life Out There. Dr. Martin Rees, a cosmologist at the University of Cambridge in England, used an e-mail message to attribute the change partly to growing scientific interest in extraterrestrial biology and the origins of life, as well as, perhaps, "the growing visibility and manifest professionalism of the SETI Institute."

The issue is so important, Dr. Rees said, that "even though the chances of success are exceedingly low, it's worth a moderate effort."

Dr. Michael Turner, an astrophysicist at the University of Chicago, used a football analogy to describe the odds, saying, "SETI is definitely throwing deep, and oh what a touchdown it would be!"

Astronomers who testified two years ago to a House subcommittee on space and aeronautics reported that members seemed to support SETI. According to the journal Nature, Representative Lamar Smith, the Texas Republican who called the hearing, said the discovery of life elsewhere in the universe would be "one of the most astounding discoveries in human history."

"Funding should match public interest," Mr. Smith said, "and I don't believe it does."

In recent years, the reports from the National Academy of Sciences have endorsed the idea of SETI, and the institute itself, once advised to change its name, has become a respected "brand" in astrobiology, said Dr. Drake, as evidenced by the recent announcement. "All of this is indeed a major sea change," he said.

It was Dr. Drake who in 1960 first pointed a radio telescope at two stars, hoping to hear the cosmic equivalent of "hi there." He did not hear anything, but he earned a curious sort of scientific immortality for his frustration.

No amount of cosmic silence has been able to discourage astronomers who theorize that radio signals can bridge the unbridgeable gulfs between stars much more cheaply than spacecraft, allowing distant species to communicate by a sort of cosmic ham radio. The Milky Way has 200 billion stars, the astronomers point out, and billions of frequencies available for signaling, if "they" exist.

SETI had been in eclipse at NASA since September 1993, when Congress, fearing a backlash if it spent tax dollars on "little green men," amended the NASA appropriation to kill a 10 year search program that had begun a year before.

Part of the project, a survey of 1,000 nearby Sun-like stars, was to have been managed by NASA's Ames Research Center in Moffett Field, Calif., and carried out by the SETI Institute, formed in 1984 as a conduit for scientists to obtain grants and conduct astrobiology research. When NASA pulled out, the astronomers at the SETI Institute decided to take the search private.

In 1995, with grants from Silicon Valley titans like David Packard, William Hewlett and Dr. Barney Oliver of Hewlett-Packard; Gordon Moore, cofounder of Intel; and Paul G. Allen, cofounder of Microsoft; as well as Arthur C. Clarke, the science fiction author and inventor of the communications satellite, the search was reborn as Project Phoenix.

The astronomers expect to finish surveying the original list in the fall, but they are already laying plans for an expanded survey of up to a million stars.

The survey will be performed by a dedicated array of 350 small radio telescopes that will be built in conjunction with the University of California at Berkeley, at its Hat Creek Observatory near Mount Lassen in Northern California. The telescope, which can be used for SETI and regular radio astronomy, will be known as the Allen Telescope Array, after Mr. Allen, who invested $11.5 million for developing the telescopes.

Meanwhile, the SETI Institute, with about 120 employees and an annual budget of $10 million, not counting the cost of the new telescope array, has grown into a powerhouse of astrobiological research.

Its scientists say they have never had any trouble obtaining support from NASA and the National Science Foundation for this branch of their activities, which are as diverse as studying the chemistry of interstellar clouds and ways to handle Martian soil samples.

In all, the institute has more than 36 individual grants and cooperative agreements with NASA, said Thomas Pierson, chairman of the institute.

"Only our SETI work is (has been, actually) without federal support for the past 10 years," Mr. Pierson wrote in an e-mail message.

That exclusion has particularly stung, as NASA has embarked on highly publicized programs to search for cosmic origins—of life, matter and everything else—and begun planning for the Terrestrial Planet Finder, which will search for Earth-like planets.

The search for intelligent life in the cosmos was a logical part of those endeavors, the staff argued, and should be eligible for federal money.

"The questions SETI asks are a natural component of the questions that get asked in astrobiology," said Dr. Christopher F. Chyba, a Stanford professor who is head of astrobiology research at the SETI Institute.

As an indication of changing fortunes, astronomers point to remarks that Dr. Weiler made at the House subcommittee hearing on July 12, 2001. Dr. Weiler made clear in the session that the Congressional ban on SETI applied just to 1994.

"NASA is no longer prohibited by any congressional language from considering funding SETI research," he said, "so SETI is currently eligible and considered fairly under peer review for NASA opportunities."

The next year, a National Academy of Sciences report, "Life in the Universe, an Assessment of U.S. and International Programs in Astrobiology," called the SETI Institute a "unique endeavor" and an "important national resource in astrobiology."

The search for intelligent life, it said, is "the most romantic and publicly accessible aspect of the search for life, yet is perhaps the most problematic."

"It would be dissembling," the report added, "to say the least, to discourage such a search (especially one enabled by private funding) at the same time that astrobiology as a whole taps into the same emotions and aspirations to excite the public about the general search for life's origins, evolution and cosmic ubiquity."

Last year, the idea of searching for intelligent signals was explicitly endorsed in the newest version of "The NASA Astrobiology Roadmap," an outline of questions and goals assembled by 200 astrobiologists inside and outside NASA as a guide to research.

Although technology may be rare in the universe, its effects may nevertheless be easier to detect from a distance than biological ones.

"Accordingly," the roadmap said, "current methods should be further developed and novel methods should be identified for detecting electromagnetic radiation or other diagnostic artifacts that indicate remote technological civilizations."

The selection of the SETI Institute as part of the Astrobiology Institute is more evidence, Dr. Chyba said, that "there has been a sea change in attitudes toward SETI, as evidenced by the 'Astrobiology Roadmap' and, explicitly, by NASA."

The higher end of evolution, intelligence, has been missing from astrobiology, he said, adding, "We're bringing it to the institute now for the first time."

In a statement from the institute, Dr. Jill Tarter, a radio astronomer who directs the search program, called the selection "the stamp of approval that what we started so long ago is a really good idea."

"The cross-fertilization of all these disciplines," Dr. Tarter said, "pays big dividends."

IN REVIEW

1. Briefly describe the funding history of the SETI Institute.
2. Why has SETI research been controversial? How have attitudes toward this research changed in recent years?
3. What is the Allen Telescope Array, and what will it be used for?
4. A Congressman is quoted in the article as saying, "Funding should match public interest." Explain what he means in the context of SETI research. Do you think his statement offers good guidance for scientific funding decisions in general? Why or why not?
5. In your opinion, should the government fund SETI research? Would your opinion change if we found microbial life on Mars? Explain.